ADVANCED COURSES OF MATHEMATICAL ANALYSIS II

EDITORS

M. V. VELASCO

A. RODRÍGUEZ-PALACIOS

Universidad de Granada, Spain

PROCEEDINGS OF THE SECOND INTERNATIONAL SCHOOL

ADVANCED COURSES OF MATHEMATICAL ANALYSIS II

Granada, Spain 20–24 September 2004

World Scientific

NEW JERSEY • LONDON • SINGAPORE • BEIJING • SHANGHAI • HONG KONG • TAIPEI • CHENNAI

Published by

World Scientific Publishing Co. Pte. Ltd.
5 Toh Tuck Link, Singapore 596224
USA office: 27 Warren Street, Suite 401-402, Hackensack, NJ 07601
UK office: 57 Shelton Street, Covent Garden, London WC2H 9HE

British Library Cataloguing-in-Publication Data
A catalogue record for this book is available from the British Library.

ADVANCED COURSES OF MATHEMATICAL ANALYSIS II
Proceedings of the 2nd International School

ISBN-13 978-981-256-652-2
ISBN-10 981-256-652-X

Printed in Singapore by World Scientific Printers (S) Pte Ltd

PREFACE

The **Second International Conference on Mathematical Analysis in Andalucía**[1] was held in Granada in the period September 20–24, 2004. With the support of both the Spanish National Government and the Andalusian Regional Government,[2] representatives of all Andalusian universities made a concerted effort to organize a meeting that provides a thorough overview of current trends and advances in the field of Mathematical Analysis. The friendly cooperation of the Andalusian research groups in this area has made possible the continuation of a project that originated in Cadiz in 2002, mainly thanks to the initiative of Antonio Aizpuru Tomás and Fernando León Saavedra. Furthermore, it is our pleasure to announce that the next conference in this series will take place in La Rábida (Huelva, Spain) in September 2007. It will be organized by the Universities of Huelva and Seville, under the direction of Tomás Domínguez Benavides and Cándido Piñeiro Gómez.

[1] HONORARY COMMITTEE

- Excmo Sr. D. Manuel Chaves González (Presidente de la Junta de Andalucía)
- Excmo Sr. D. Francisco Vallejo Serrano (Consejero de Innovación Ciencia y Empresa de la Junta de Andalucía)
- Excmo Sr. D. David Aguilar Peña (Rector Magní fico de la Univ. de Granada)
- Excmo Sr. D. Alfredo Martínez Almécija (Rector Magnífico de la Univ. de Almería)
- Excmo Sr. D. Diego Sales Márquez (Rector Magní fico de la Univ. de Cádiz)
- Excmo Sr. D. Eugenio Martínez Vilches (Rector Magn ífico de la Univ. de Córdoba)
- Excmo Sr. D. Antonio Ramírez Verger (Rector Magn ífico de la Univ. de Huelva)
- Excmo Sr. D. Luis Parra Guijosa (Rector Magnífico de la Univ. de Jaén)
- Excma Sra. Dña. Adelaida de la Calle Martín (Rectora Magníf. de la Univ. de Málaga)
- Excmo Sr. D. Miguel Florencio Lora (Rector Magní fico de la Univ. de Sevilla)

MAIN ORGANIZER: M. Victoria Velasco Collado (Univ. de Granada)
SCIENTIFIC/ORGANIZING COMMITTEE:

- University of Granada: Juan Francisco Mena Jurado, Rafael Payá Albert, Ángel Rodríguez Palacios.
- University of Almería: Amin el Kaidi, Juan Carlos Navarro Pascual.
- University of Cádiz: Antonio Aizpuru Tomás, Fernando León Saavedra.
- University of Córdoba: Juan Carlos Díaz Alcaide.
- University of Huelva: Cándido Piñeiro Gó mez, Ramón Rodríguez Álvarez.
- University of Jaén: Miguel Marano Calzolari, Francisco Roca Rodríguez.
- University of Málaga: Antonio Fernández Ló pez, Daniel Girela Álvarez, Francisco Javier Martín Reyes.
- University of Sevilla: Santiago Díaz Madrigal, Tomás Domínguez Benavides, Carlos Pérez Moreno, Luis Rodrí guez Piazza.

LOCAL COMMITTEE: M. Dolores Acosta Vigil, Julio Becerra Guerrero, Antonio Moreno Galindo, Antonio Peralta Pereira.

[2] Acción Especial BFM2002-11783-E and Acción Coordinada ACC1273-FQM-2003 were specific grants

The goal of these conferences is to provide an opportunity for distinguished researchers to explore a variety of topics, to initiate the discussion of open problems, and to present tools and techniques that might prove effective in their resolution. The Second Conference consisted of seminars and one-hour plenary talks. While the one-hour talks were intended to provide an overview of a variety of areas of contemporary interest, the seminars extended over several days and permitted in-depth discussion on certain specific topics. Moreover, all participants had the opportunity to present new results of their own research in short communications.

The participation and collaboration of outstanding researchers in the 2004 Conference surpassed all of our expectations and predictions. Approximately 80 lectures were given on topics of current scientific interest, more than meeting our goals. We would like to express our most sincere gratitude to all of those who, through their work, presence, and scientific contributions, made the success of this Conference and the publication of these proceedings possible.

The present book is comprised of works that are closely related to the content of the seminars and plenary talks given at the 2004 Conference. It includes contributions from Richard M. Aron, Fernando Bombal, Javier Duoandikoetxea, Gilles Godefroy, Nigel J. Kalton, Lawrence Narici, Michael M. Neumann, Baldomero Rubio, Manuel Valdivia, Joan Verdera, and Felipe Zó (co-authored with Héctor H. Cuenya).

We would like to take the opportunity of this preface to pay special tribute to the memory of Miguel de Guzmán, who was scheduled to be one of our guest speakers, but passed away shortly before the holding of the Conference. We know we have been deprived of what would surely have been an outstanding contribution. The paper of Baldomero Rubio contains a touching biographical sketch of Miguel de Guzmán, and addresses interesting ideas about the teaching of mathematics, a topic of great importance in which Miguel de Guzmán was the leading authority in Spain.

The principal results of the contribution by Manuel Valdivia, certainly the dean of our main speakers, have not been published before. These results are related to the study of the so-called "lineable" and "spaceable" subspaces of a locally convex space. The article culminates in the construction of a subspace of the space $H(\Omega)$ (of all holomorphic functions on a domain Ω of $\mathbb{C}^n$) which is dense in $H(\Omega)$ for the compact open topology, is ordered nearly-Baire, and has the property that each of its elements does not extend holomorphically outside Ω. The paper also includes a deep general study of nearly-Baire and ordered nearly-Baire locally convex spaces.

The paper by Fernando Bombal reviews much of the literature about the work of Alexander Grothendieck and provides the reader with a perspective of his work on functional analysis together with a glimpse of certain parts of his heritage. The paper focuses mainly on Grothendieck's contributions to the theory of topological tensor products, first in the general setting of locally convex spaces (the subject of his Ph.D. dissertation) and later as a powerful tool to study the structure of Banach spaces (mainly contained in the famous Sao Paolo's *Résumé*). In addition to providing a careful review of these two masterpieces, Bombal comments on Grothendieck's remarkable paper on $C(K)$-spaces in the *Canadian Journal of Mathematics.*

The article by Felipe Zó and Héctor Cuenya consists mainly of previously unpublished results. The authors consider a family of monotone function seminorms acting on Lebesgue measurable functions $F : B \subset \mathbb{R}^n \to \mathbb{R}^k$, where $B := \{x \in \mathbb{R}^n : |x| \leq 1\}$, as well as the best approximation function $P_\varepsilon \in A$, where A is a given subspace of polynomials. They study the asymptotic behavior of a normalized error function as well as the limit of the net $\{P_\varepsilon\}$ as $\varepsilon \to 0$. This vector valued approach extends and unifies some classical problems in best local approximation theory.

The contributions of Richard M. Aron, Javier Duoandikoetxea, Gilles Godefroy, Nigel J. Kalton, Lawrence Narici, Michael M. Neumann, and Joan Verdera are survey papers on the respective topics of their specialization.

The paper of Richard M. Aron deals with "lineability", i.e. the problem to decide when vectors having some (perhaps strange) property $\mathcal{P}$ form an infinite-dimensional space. The author concentrates on three very different instances of lineability: zeros of polynomials, hypercyclic vectors associated to hypercyclic operators, and differentiability of continuous functions. Positive and negative results are given in each instance, and many results (both by the author and by others) are described. The paper concludes with a comment on the relation between lineability and the Baire category argument. A good number of well-chosen examples and many references are provided.

The contribution by Javier Duoandikoetxea about the Hardy–Littlewood maximal function and some related topics represents an excellent overview that will be appreciated not only by the experts, bur also by graduate students and other non-specialists who are interested in an introduction to this important and very useful topic of harmonic analysis. The author introduces the basic variations on the classical maximal operator and explains the connection with the differentiation theorem. Some of the important ideas behind the proofs are sketched. The author also discusses two current research areas (and some of the open problems) that have arisen from the study of this operator: the Kakeya maximal operator and maximal operators defined by sets of directions. An extensive bibliography on the topic is given.

The article by Gilles Godefroy deals with hypercyclic operators, focusing on some important new results introduced by Sophie Grivaux and Frédéric Bayart in a series of papers. The author presents the theorems and their proofs, avoiding technicalities whenever possible. In this way, this paper becomes an invitation to further reading, an introduction to the original articles, and a preparation for the challenging questions which remain open.

The paper by Nigel Kalton surveys recent results (some of them from the author) concerning greedy, quasi-greedy, and almost greedy bases for Banach spaces. The paper provides a gentle introduction to this theory.

The informative article by Lawrence Narici is devoted to presenting the origins, history, and evolution of what is known as the Hahn–Banach theorem, and more precisely, its analytic version. The author reviews some of the direct antecedents of this theorem: the study of infinite systems of linear equations, and the moment problem. The fundamental contributions of F. Riesz and E. Helly, as well as the work of S. Banach and H. Hahn on the subject, are analyzed in detail. Also surveyed are extendible Banach spaces, i.e., those Banach spaces that can play the role of the scalar field in the classical version of the Hahn–Banach theorem for normed spaces. Moreover, the paper gives an interesting account of results about the uniqueness of Hahn–Banach extensions and non-Archimedean versions of the Hahn–Banach theorem.

The contribution by Michael M. Neumann nicely details the history and origins of the classical Cesàro operator, before proceeding to a study of the fine structure of the spectrum and the spectral decomposition properties (specifically, subnormality and hyponormality) of generalized Cesàro operators with rational symbol on Hardy and weighted Bergman spaces. While mainly a survey article, the paper also contains some interesting new contributions.

The paper by Joan Verdera reviews the basics of classical potential theory in such a way that the formal analogy between analytic capacity and classical Wiener capacity becomes evident, and the semi-additivity problem becomes natural. This long-standing open problem in complex analysis has been solved recently using real variable methods.

Most of the aforementioned survey papers will be authoritative references for the optimal version of results scattered in the literature through many separate papers. The editors would like to express their gratitude to the authors listed above for their careful preparation of their contributions to this book. The editors are also very grateful to the referees of the papers contained in this book, who must remain anonymous, but who donated their time generously and selflessly.

Last but not least, we would like to thank the representatives of all of the public institutions that supported the Second International Conference on Mathematical Analysis in Andalusia from the beginning. In particular, we would like to thank the Regional Counselor for Innovation, Science and Business, Francisco Vallejo Serrano, and the Chancellor of Granada University, David Aguilar Peña, both of whom granted us the honor of opening what started as a scientific meeting and ends with the publication of the present book.

Granada, Autumn 2006

The Editors

CONTENTS

Preface v

Linearity in Non-Linear Situations
R. M. Aron 1

Alexander Grothendieck's Work on Functional Analysis
F. Bombal 16

The Hardy–Littlewood Maximal Function and Some of Its Variants
J. Duoandikoetxea 37

Linear Dynamics
G. Godefroy 57

Greedy Algorithms and Bases from the Point of View of Banach Space Theory
N. J. Kalton 76

On the Hahn–Banach Theorem
L. Narici 87

Spectral Properties of Cesàro-like Operators
M. M. Neumann 123

Tribute to Miguel de Guzmán: Reflections on Matematical Education Centered on the Mathematical Analysis
B. Rubio Segovia 141

On Certain Spaces of Holomorphic Functions
M. Valdivia 151

Classical Potential Theory and Analytic Capacity
J. Verdera 174

Best Approximations on Small Regions — A General Approach
F. Zó and H. H. Cuenya 193

LINEARITY IN NON-LINEAR SITUATIONS

RICHARD M. ARON
Mathematics Department
Kent State University
Kent, Ohio 44242
USA
aron@math.kent.edu

We give an informal account of recent progress and questions related to the general problem: When does the subset of a vector space that has a certain (interesting) property contain an infinite dimensional vector space?

In 2001, D. García, M. Maestre, and the author published an expository article [4] in which we introduced the terms *lineable* and *spaceable*. The aim of these lectures is to give an accessible description of these notions which, we hope, will encourage others to find many more new and interesting instances of lineability and spaceability.

As the reader will soon observe, these notes will be leisurely, with either brief indications of some proofs or else no proofs at all. To highlight our casual approach, we "officially" begin with not one, but two introductions.

1. Two introductions

Introduction 1. Let's first recall three classical results.

(a). (Weierstrass ($\pm$ 1856)) There is a continuous, nowhere differentiable function $f : \mathbb{R} \to \mathbb{R}$.

(b). (Lebesgue (1904)) There is a function $f : \mathbb{R} \to \mathbb{R}$ with the property that for any $a < b$, the restriction $f|_{(a,b)} : (a, b) \to \mathbb{R}$ is surjective. (Such a function f is called *everywhere surjective.*)

(c). (Birkhoff (1929)) There is an entire function $f : \mathbb{C} \to \mathbb{C}$ such that the set of translated functions $\{f(z), f(z+1), f(z+2), ...\}$ is dense in the space $\mathcal{H}(\mathbb{C})$ of entire functions, with the usual compact-open topology.

To us, the mere existence of such functions is surprising. In fact, there are very many such functions and, indeed, there are large vector spaces of such functions! To continue, it will be very helpful to have two new terms (which, until fairly recently, did not exist in the English language).

Definition 1 Let E be a Banach or Fréchet space, let $\mathcal{P}$ be a property that elements of E might have, and let $S \equiv \{x \in E \; : \; x \text{ has property } \mathcal{P}\}$. We say that $\mathcal{P}$ is *lineable* if $S \cup \{0\}$ contains an infinite dimensional vector space, and that S is *spaceable* if $S \cup \{0\}$ contains a *closed* infinite dimensional vector space.

(i) **Theorem** [21] The property of being continuous and nowhere differentiable is a spaceable property in the space $\mathcal{C}([0,1])$ of continuous real-valued functions on $[0,1]$.
(ii) **Theorem** [6] The property of being everywhere surjective is lineable in the space of all functions $f : \mathbb{R} \to \mathbb{R}$. (Note that spaceability is not an issue here, since there is no natural topology on the space of all functions on $\mathbb{R}$.)
(iii) **Theorem** [14] The property of being an entire function f whose set of translates

$$\{f(z), f(z+1), ..., f(z+n), ...\}$$

is dense in $\mathcal{H}(\mathbb{C})$ is spaceable.

It is natural to ask for criteria for when lineability occurs, and when a property is found to be lineable it is natural to wonder whether it is also spaceable. By and large, these natural questions are open and in many situations they seem to be quite difficult. This leads us to our second introduction.

Introduction 2. To us, it seems that very often when one or the other of lineability and spaceability is proved, the result is surprising and almost magical. Thus, a second personal motivation for our interest in this topic, is the following: How does one get *intuition* about whether a property $\mathcal{P}$ is lineable, spaceable, or neither? The reader will note that despite the abundance of examples that are presented, we are still confounded by this problem of how to develop a sense of what is true.

Let us illustrate the problems we confront with a number of examples.

Examples 1. Let $X = \mathcal{C}[0,1]$, and let $\mathcal{P}$ be the property: $f \in X$ *attains its maximum at precisely one point of* $[0,1]$. Is $\mathcal{P}$ a lineable, or a spaceable, property? That is, does the set $\{f \in \mathcal{C}[0,1] \ : \ f$ attains its norm at precisely one point $\} \cup \{0\}$ contain an infinite dimensional vector space; does it contain a closed infinite dimensional vector space?

2. Now let $X = \{f \in \mathcal{C}(\mathbb{R}) \ : \ |f(x)| \to 0 \ \text{ as } x \to \infty\}$, which is a Banach space with the max-norm, and let $\mathcal{P}$ be the same property as in **1** above: f satisfies $\mathcal{P}$ if and only if f attains its maximum at exactly one point of $\mathbb{R}$. Is this a lineable property: is it spaceable?

3. Let $T \cong \mathbb{R}/2\pi$ be the unit circle. Let $\mathcal{P}$ be the property that the Fourier series of a function $f \in \mathcal{C}(T)$ *does not* converge at, say, the point $z = 1$. Is $\mathcal{P}$ a lineable, or a spaceable, property?

4. Let $\mathcal{P}$ be the property that a function $f \in \mathcal{C}[0,1]$ is everywhere differentiable. Is $\mathcal{P}$ lineable? Is it spaceable?

5. Fix a non-empty open subset $U \subset \mathbb{C}^n$, and let $\mathcal{H}(U)$ denote the holomorphic (complex analytic) functions on U, endowed with the standard Fréchet topology of uniform convergence on compact subsets of U. Let $\mathcal{P}$ be the property that a function $f \in \mathcal{H}(U)$ cannot be continued beyond any point of the boundary of U. (In other words, U is the *domain of existence* of f.) Lineable? Spaceable?

6. Fix an infinite dimensional Banach space E, and let $\mathcal{P}$ be the property that a continuous linear functional $\varphi \in E^*$ attains its norm; that is $||\varphi|| = \phi(x_0)$ for some $x_0 \in E$, $||x_0|| = 1$. By a classical result of R. C. James [27], E is reflexive if and only if every such φ is norm-attaining. On the other hand, in [16], E. Bishop and R. Phelps showed that for every E, the set of norm-attaining functionals in E^* is a norm-dense subset of E^*. Is $\mathcal{P}$ a lineable property; is it a spaceable property?

7. Let $T_2 : \ell_2 \to \ell_2$ be the Rolewicz continuous linear operator given by $T_2(x_1, x_2, ...) = 2(x_2, x_3, ...)$ [31]. Consider the set $\mathcal{HC}(T_2)$ of vectors $x = (x_n) \in \ell_2$ whose orbits under T is dense in ℓ_2. (In other words, $\overline{Orb(T,x)} = \overline{\{x, T_2 x, T_2^2 x, ..., T_2^n x, ...\}} = \ell_2$.) Does $\mathcal{HC}(T_2) \cup \{0\}$ contain an infinite dimensional subspace of ℓ_2? Does it contain a closed, infinite dimensional subspace of ℓ_2?

8. Let $T : \mathcal{H}(\mathbb{C}) \to \mathcal{H}(\mathbb{C})$ be a differential operator acting on the space $\mathcal{H}(\mathbb{C})$ of entire functions. Let $\mathcal{P}$ be the property that the orbit of an entire function f under T is dense in $\mathcal{H}(\mathbb{C})$. Is $\mathcal{P}$ lineable; spaceable?

9. Let E be an infinite dimensional Banach space over $\mathbb{K} = \mathbb{R}$ or $\mathbb{C}$, and let $P : E \to \mathbb{K}$ be a continuous polynomial. Let $\mathcal{P}$ be the property that a vector $x \in E$ is such that $P(x) = P(0)$. Is this a lineable property? In other words, does the set $\{x \in E \ : \ P(x) = P(0)\}$ contain an infinite dimensional vector space?

Note that if this set contains an infinite dimensional vector space, then it automatically contains a closed infinite dimensional vector space; in other words, in this case, lineability and spaceability are the same problem. The next section will begin by recalling the necessary background, including relevant definitions, with a review of the current status of knowledge on this and the following finite dimensional version of the same problem.

10. Consider a polynomial in n real or complex variables, $P : \mathbb{K}^n \to \mathbb{K}$ such that $P(0) = 0$. Consider the zero set $P^{-1}(0)$ of P. Does $P^{-1}(0)$ contain a vector space? If so, what can we say about the dimension of $P^{-1}(0)$?

What is especially intriguing concerning these ten problems (and many others) is the near total lack of accurate intuitive insight that we possess about them! In the rest of the paper, we will give some answers that are incomplete in many cases, in hopes that the reader will continue this very incomplete study.

2. Lineability of the zeros of polynomials

Our aim here is to examine problems 9 and 10 above. We will review some non-technical arguments that show that the situation for complex polynomials is reasonably well understood. We will briefly discuss the real case, which is considerably more challenging. Finally, we will describe some open problems for both the real and complex situations.

The Complex Case

Definition 2. Let E be a Banach space over $\mathbb{K}$, $k \in \mathbb{N}$. A continuous function $P : E \to \mathbb{K}$ is said to be a *k–homogeneous polynomial* if there exists a k–linear continuous, symmetric functional $A : E \times \cdots \times E \to \mathbb{K}$ such that $P(x) = A(x, ..., x)$ for all $x \in E$. (In this case, the associated symmetric functional A is necessarily unique.) We say that P is a *polynomial of degree k* if P is a finite sum of j–homogeneous polynomials P_j, $P = P_0 + P_1 + \cdots + P_k$.

Examples. We agree that a 0–homogeneous polynomial is a constant. A 1–homogeneous polynomial is just a linear form. A 2–homogeneous polynomial $P : \mathbb{K}^n \to \mathbb{K}$ corresponds to a *quadratic form* Q (i.e. a symmetric $n \times n$ matrix over $\mathbb{K}$), via $P(x) = x^t Q x \quad (x \in \mathbb{K}^n)$, which in turn corresponds to the symmetric bilinear form A on $\mathbb{K}^n \times \mathbb{K}^n$, $\quad A(x, y) = y^t A x$.

To simplify the statements of the results that follow, we will assume from now on all that polynomials considered will have value 0 at 0. One of the most important results to date in this area is the following theorem due to A. Plichko and A. Zagorodonyuk.

Theorem 3. [30] Let E be an infinite dimensional vector space over $\mathbb{K} = \mathbb{C}$, and let $P : E \to \mathbb{C}$ be a polynomial. Then $P^{-1}(0)$ contains an infinite dimensional subspace.

A finite dimensional version of this result may have independent interest.

Corollary 4. [10] Given $m, d \in \mathbb{N}$, there is $k = k(m, d) \in \mathbb{N}$ such that if $P : \mathbb{C}^k \to \mathbb{C}$ is a polynomial of degree d, then $P^{-1}(0)$ contains an m–dimensional subspace.

In other words, if we want to be sure that $P^{-1}(0)$ contains a vector space of (at least) a certain dimension m for all polynomials of a certain degree d, then we can find an integer k such that every polynomial of degree at most d in k complex variables will vanish on a subspace of dimension at least m. Or, described in still another way, given d, as the number k of variables tends to infinity, the dimension of the subspace contained in the zero set of a polynomial on $\mathbb{C}^k$, whose degree is at most d also tends to infinity.

We now sketch a proof of these results, following [10] (which, after all, is based on [30]). For it, we will use one well-known fact from several complex variables: *Fact:* If $f : \mathbb{C}^n \to \mathbb{C}$ is an analytic function, where $n \geq 2$, then either $f^{-1}(0) = \emptyset$ or $f^{-1}(0)$ is an unbounded set. We will apply this to the case when f is a non-constant polynomial (in fact, in the case of a polynomial in several variables, the 'fact' is easy to show), and thus we will be able to conclude that $f(z) = 0$ for some $z \neq 0$.

Proof idea for Theorem 3 and Corollary 4. The formal proof is an inductive argument on the degree of homogeneity d. The constructive idea of this can already be seen in the case $d = 1$ and $d = 2$. If $d = 1$, then P is just a continuous linear form whose kernel is a hyperplane. (In other words, if P is a linear form in k variables, the dimension of the zero set of P is $m = k - 1$.) Now, let $P : E \to \mathbb{C}$ be a $2-$homogeneous polynomial with associated symmetric bilinear form A. If $\dim E \geq 2$, then by our fact, there is a non-zero vector $x_1 \in E$ such that $P(x_1) = 0$. Next, consider the set $S_1 \equiv \{x \in E \ : \ A(x, x_1) = 0\}$. This set, which in fact is a hyperplane, contains x_1. Thus S_1 can be written as $E_1 \bigoplus [x_1]$, where $\dim E_1$ is 2 less than $\dim E$. (Of course, this statement is not tremendously profound if $\dim E = \infty$.) If $\dim E_1 \geq 2$, then our fact will yield a second non-zero vector $x_2 \in E_1$ for which $P(x_2) = 0$. What this really means is that if $\dim E \geq 4$, then we can find independent vectors x_1, x_2 such that $P(x_1) = P(x_2) = 0$. Note that in addition $[x_1, x_2] \subset P^{-1}(0)$. To see this, it is enough to check that $x_1 + x_2 \in P^{-1}(0)$:

$$P(x_1+x_2) = A(x_1+x_2, x_1+x_2) = A(x_1, x_1)+A(x_1, x_2)+A(x_2, x_1)+A(x_2, x_2),$$

which since A is symmetric,

$$= P(x_1) + 2A(x_2, x_1) + P(x_2) = 0,$$

since $x_2 \in E_1$. Continuing, let $S_2 \equiv \{x \in E_1 \ : \ A(x, x_2) = 0\}$, so $S_2 = E_2 \bigoplus [x_2]$, where $\dim E_2 = \dim E_1 - 2$. Once again, if $\dim E_2 \geq 2$, then we can find $x_3 \in E_2, x_3 \neq 0$, such that $P(x_3) = 0$. The general argument proceeds by a not too difficult induction.

Two observations should be made at this point, both dealing with the method of proof and both having 'resonance' with the real case that we'll describe, very briefly, in a few paragraphs.

Remarks 5.1. The argument is essentially *constructive.* In particular, to 'build' a subspace of dimension k contained in $P^{-1}(0)$ one can use a previously constructed subspace of dimension $k - 1$ and add an appropriate vector to it.

5.2. The argument is *countable.* To us, it seems natural to expect that if $\dim E$ is uncountable, then $P^{-1}(0)$ should contain a non-separable space. Our proof merely shows that there is a separable space contained in $P^{-1}(0)$. Some interesting recent work by Plichko and Zagorodnyuk [11] has been done on this problem, which suggests that the phrase "natural to expect" in the previous sentence may not be true. On the other hand, a number of very interesting positive results have recently been obtained by M. Fernández [20] and C. Soares [34].

The Real Case

We now study the size of vector spaces contained in $P^{-1}(0)$ in the case of polynomials $P : E \to \mathbb{R}$, $P(0) = 0$. Of course, $\sum_{j=1}^{n} x_j^2 : \mathbb{R}^n \to \mathbb{R}$ shows that no extremely general result is possible. However, we can accomplish a reasonable amount. The first result is just an easy exercise in linear algebra.

Proposition 6. (case $d = 2$) Let $P : \mathbb{R}^k \to \mathbb{R}$ be a $2-$homogeneous polynomial and let Q be the associated quadratic form. Let $p =$ the number of positive eigenvalues of Q, let $n =$ the number of negative eigenvalues of Q, and let $z =$ the number of 0 eigenvalues of Q. Then $P^{-1}(0)$ contains a subspace of dimension $= \min\{p, n\} + z$.

It would be interesting to know if there is an analogous result for the case of $d = 4-$homogeneous polynomials.

For the rest of this section, we will focus on homogeneous polynomials of odd degree. Using a fairly technical argument that (apparently) works only for $d = 3$, R. Gonzalo, A. Zagorodnyuk, and the author were able to show that the analogue of Corollary 4 holds [5]. This was later extended by P. Hajek and the author (using a different technical, counting argument), as follows:

Theorem 7. [7] Given $m, d \in \mathbb{N}$ where d is odd, there is $k = k(m, d) \in \mathbb{N}$ such that if $P : \mathbb{R}^k \to \mathbb{R}$ is an odd polynomial of degree d, then $P^{-1}(0)$ contains an $m-$dimensional subspace.

Remark 8. The proofs in both [5] and [7] are *non-constructive.* Knowing the $k-1$ vectors that span a subspace of $P^{-1}(0)$ does not help us to find a k dimensional subspace. Consequently, we don't know whether the analogue of Plichko's and Zagorodnyuk's Theorem 3 holds. Specifically, suppose that $P : E \to \mathbb{R}$ is an odd-homogeneous polynomial on an infinite dimensional real Banach space E. Although we know that $P^{-1}(0)$ contains a vector space of every finite dimension, we do not know whether $P^{-1}(0)$ contain an infinite dimensional vector space.

3. Non-spaceability of differentiable functions

As was seen in §2 and will be observed in later sections, in many situations sets that might otherwise be said to have pathological, non-linear properties contain a rich linear structure. In this section, we will mention a counterpart to this situation. Namely, we will describe how poor the structure can be in the case of functions with what most people would call very good properties. Namely, we will outline the proof of V. Gurariy and W. Lusky [25] of the following theorem, due originally to Fonf, V. Gurariy, and V. Kadets [21].

Theorem 9. Let $V \subset \mathcal{C}[0,1]$ be a closed subspace of $\mathcal{C}[0,1]$ consisting of everywhere differentiable functions. Then $\dim V < \infty$.

Of course, there are infinite dimensional vector spaces of everywhere differentiable (in fact, $\mathcal{C}^\infty$) functions contained in $\mathcal{C}[0,1]$. The point is that none of these spaces is complete.

The basic concept behind the following proof is well known; see, for example, ([32], 7.18). For the proof, we will need some terminology. First, a sequence (A_k) of subsets of $[0,1]$ is said to be *condensed* if there is some sequence (ϵ_k) of positive numbers that tends to 0 such that each A_k is an ϵ_k−net for $[0,1]$. That is, for all $t, 0 \le t \le 1$, there is $s \in A_k$ such that $|s-t| < \epsilon_k$. A sequence of functions $(f_k) \subset \mathcal{C}[0,1]$ is said to be *condensed* if the associated sequence of zero sets, $(f_k^{-1}(0))$, is condensed.

We will omit the proof of the following slightly technical, but not difficult, lemma.

Lemma 10. Let $V \subset \mathcal{C}[0,1]$ be an infinite dimensional space. Then V contains a condensed sequence of functions.

Proof of Theorem 9. Suppose that there is an infinite dimensional Banach space $V \subset \mathcal{C}[0,1]$ all of whose elements are everywhere differentiable. By Lemma 10, there is a condensed sequence $(f_k) \subset V$, and we may assume that all the f_k have norm one. Let $A_k = f_k^{-1}(0)$ and let $t_k \in [0,1]$ be such that $|f_k(t_k)| = 1$. By passing to a subsequence, we may suppose that $(t_k) \to t_0$ for some $t_0 \in [0,1]$. Let $s_k \in A_k$ be close to t_k and set

$$t_k^* \equiv \begin{cases} t_k \text{ if } |f_k(t_0)| \leq \frac{1}{2} \\ s_k \text{ if } |f_k(t_0)| > \frac{1}{2} \end{cases}.$$

Because of the condensed property, we may assume that

$$|t_k^* - t_0| < \frac{1}{32^k}.$$

Note also that $|f_k(t_k^*) - f_k(t_0)| \geq \frac{1}{2}$. Now define $f : [0,1] \to \mathbb{R}$ by

$$f(t) = \sum_{j=1}^{\infty} \frac{\theta_j}{16^j} f_j,$$

where we have chosen $\theta_j = \pm 1$ so that $\sum_{j=1}^{k-1} \frac{\theta_j}{16^j}(f_j(t_k^*) - f_j(t_0))$ and $\sum_{j=1}^{k} \frac{\theta_j}{16^j}(f_j(t_k^*) - f_j(t_0))$ have the same sign. It is easy to see that $f \in V$. However, f is not differentiable at t_0. Indeed,

$$|f(t_k^*) - f(t_0)| \geq \frac{1}{16^k}|f_k(t_k^*) - f_k(t_0)| - \sum_{j=k+1}^{\infty} \frac{1}{2^j}|f_j(t_k^*) - f_j(t_0)| \geq \frac{1}{4 \cdot 16^k},$$

provided $k \geq 2$. Thus, $\lim_{k\to\infty} \frac{f(t_k^*) - f(t_0)}{t_k^* - t_0}$ does not exist, and we have a contradiction. Q.E.D.

4. Lineability of hypercyclic operators

Let $T : X \to X$ be a continuous linear operator on a Banach or Fréchet space X. We say that T is *hypercyclic* if there is a vector $x \in X$ whose orbit under T, $\mathrm{Orb}(T,x) = \{x, Tx, T^2x, ..., T^nx, ...\}$ is dense in X; the vector x is called a *hypercyclic vector* associated to T. Clearly, hypercyclic operators are much harder to come by than *cyclic* operators T, which have the defining property that *span* $\mathrm{Orb}(T,x)$ is dense in X. Despite this, hypercyclic operators are (perhaps) surprisingly ubiquitous, and in this section we will briefly comment on this and on the fact that the set of hypercyclic vectors x that are associated to an operator T contain vector spaces that are large and not yet fully understood.

It might be fair to say that the origin of hypercyclicity dates from the short note of G. Birkhoff [15] over 75 years ago. It was Birkhoff who first proved that the translation operator $\tau_1 : \mathcal{H}(\mathbb{C}) \to \mathcal{H}(\mathbb{C})$, given by $\tau_1(f)(z) = f(z+1) \quad (f \in \mathcal{H}(\mathbb{C}),\ z \in \mathbb{C})$, is hypercyclic. Around 25 years later, G. MacLane [28] proved that the derivative operator $D : \mathcal{H}(\mathbb{C}) \to \mathcal{H}(\mathbb{C}),\ \ D(f) = f'$, is hypercyclic. Nearly fifteen years ago, G. Godefroy and J. Shapiro [22] generalized these two results to what are called *convolution operators,* or *infinite order differential operators,* as follows:

Theorem 11. Let $T : \mathcal{H}(\mathbb{C}) \to \mathcal{H}(\mathbb{C})$ be a continuous linear operator that satisfies the following two conditions:

(i). $T \circ D = D \circ T$;

(ii). $T \neq cI$ for any $c \in \mathbb{C}$, where $I : \mathcal{H}(\mathbb{C}) \to \mathcal{H}(\mathbb{C})$ is the identity operator.

Then T is hypercyclic.

However, it is only in recent years that this area of analysis has "taken off" as a research area. We will limit ourselves to reviewing three topics: (i). The dense lineability and the spaceability of the space of hypercyclic vectors; (ii). Vectors that are hypercyclic for very many hypercyclic operators; (iii). Some open problems.

P. Bourdon was perhaps the first to make the following not difficult, but important observation.

Proposition 12. [18] If $T : X \to X$ is hypercyclic, then the set $\mathcal{HC}(T)$ of hypercyclic vectors for T contains a dense vector space.

Proof for X a complex Banach space. (The argument for Fréchet spaces and for *real* spaces is somewhat more complicated.)

First, one observes that if T is hypercyclic, then $T - \lambda I$ has dense range for any complex number λ. Indeed, if this were not the case for some λ, then by the Hahn-Banach theorem there would exist a non-zero $\phi \in X^*$ such that $\phi(T - \lambda I)(x) = 0$ for all $x \in X$. As a result, we would have that $T^t(\phi) = \lambda\phi$ for some nonzero $\phi \in X^*$. Consequently, for any $x \in X, \{\phi(T^n(x)) \mid n \in \mathbb{N}\} = \{\lambda^n(\phi(x)) \mid n \in \mathbb{N}\}$, and this set is never dense in $\mathbb{C}$. Since this contradicts the hypercyclicity of T, it follows that $T - \lambda I$ always has dense range.

Note that if x is hypercyclic for T, then $\{P(T)(x) \mid P \text{ is a polynomial }\}$ is a dense vector space in X (just take $P(T) = T^n$, $n \in \mathbb{N}$). Next, we claim that any non-zero $P(T)$ has dense range. To see this factor P into linear terms, $P(z) = a(z-\lambda_1)\cdots(z-\lambda_k)$, so that $P(T) = a(T-\lambda_1 I)\cdots(T-\lambda_k I)$. Then apply the first observation, above. Finally, $P(T)(x)$ is a hypercyclic vector for T, since $\{P(T)(x), T(P(T)(x)), ..., T^n(P(T))(x), ...\} = P(T)(\{x, T(x), ..., T^n(x), ...\})$. Q.E.D.

From this, it follows that there are no hypercyclic operators on $\mathbb{C}^n$ for any $n \in \mathbb{N}$.

Now, what spaces admit hypercyclic operators? By a result proved independently by S. Ansari [2] and L. Bernal [13], every infinite dimensional Banach space admits a hypercyclic operator. In fact, this is even true for infinite dimensional Fréchet spaces, as J. Bonet and A. Peris have observed [17]. Thus, dense lineability of hypercyclic vectors occurs in every infinite dimensional space. On the other hand, the situation concerning spaceability is much more complicated and only partially understood. For instance, Bernal-González and Montes-Rodríguez [14] proved that the set of hypercyclic entire functions for the Birkhoff translation τ_1 (together with the 0 function, as usual) contains a closed hypercyclic subspace (for the compact-open topology). On the other hand, in [29] Montes shows that the set of vectors in ℓ_2 that are hypercyclic for the Rolewicz backward shift does not contain a closed infinite dimensional subspace; that is, the set of hypercyclic vectors for the weighted backward shift $(x_1, x_2, x_3, ...) \to 2(x_2, x_3, x_4, ...)$ is not spaceable. Does the same hold for $\mathcal{HC}(D)$? In a rough sense, the operator D behaves like the Rolewicz operator, and so a first guess might be that $\mathcal{HC}(D)$ does not contain a closed subspace. But guesses are rarely publishable!

This brings us to the obvious question: How can one tell, or how can one even guess, whether the set of vectors that are hypercyclic for a given hypercyclic operator is spaceable?

To us, the mere existence of a hypercyclic operator is counter-intuitive, at least at first. However, as we just noted in (i) above, given a hypercyclic operator T, the set of vectors that are hypercyclic for T is always a dense vector space. So, our intuition notwithstanding, there seem to be plenty of hypercyclic vectors for at least standard operators. In two recent articles, even more has been shown to hold.

In [1], E. Abakumov and J. Gordon prove the following remarkable result.

Theorem 13. There is a dense G_δ of vectors lying in the set

$$\cap_{\{\lambda \in \mathbb{C} \mid |\lambda|>1\}} \mathcal{HC}(T_\lambda),$$

where $T_\lambda : \ell_2 \to \ell_2$, $T_\lambda(x_1, x_2, ...) = \lambda(x_2, x_3, ...)$. In other words, there is a dense G_δ set of vectors in $x \in \ell_2$ such that $\{x, T_\lambda(x), T_\lambda 2(x), \cdots\}$ is dense in ℓ_2 for every $\lambda \in \mathbb{C}$, $|\lambda| > 1\}$.

In addition to giving a new proof of the result in [1], G. Costakis and M. Sambarino [23] have shown that there is a dense G_δ set of entire functions contained in the set

$$\cap_{\{b \in \mathbb{C} \mid b \neq 0\}} \mathcal{HC}(\tau_b),$$

where $\tau_b : \mathcal{H}(\mathbb{C}) \to \mathcal{H}(\mathbb{C})$. $f(z) \rightsquigarrow f(z+b)$. In fact, as A. Hallack observes [26], there is even a dense vector space of such functions.

This leads us to several questions. First, in light of the result of [23] it is natural to ask for even more. Namely, is the set

$$\cap\{\mathcal{HC}(T) \mid T : \mathcal{H}(\mathbb{C}) \to \mathcal{H}(\mathbb{C}) \text{ a convolution operator, } \mathrm{T} \neq \mathrm{cI}\}$$

non-empty? A seemingly simpler question, which would give a negative answer, is the following: Given $f \in \mathcal{H}(\mathbb{C})$, is there a convolution operator T such that $T(f) = 0$? (Clearly, if $T(f) = 0$, then f cannot be hypercyclic for T.)

Next, note that the set of hypercyclic convolution operators on $\mathcal{H}(\mathbb{C})$ form (or, really, almost form) an algebra under composition. To be specific, if S and T are two hypercyclic convolution operators on $\mathcal{H}(\mathbb{C})$, then so is their composition $S \circ T$ except in the trivial case when $S \circ T = cI$ for some $c \in \mathbb{C}$. Now, in [8] (see also [19]), the question of finding hypercyclic operators on $\mathcal{H}(\mathbb{C})$ that are not convolution operators is examined. In fact, any operator of the form $T_{\lambda,b}(f)(z) = f(\lambda z + b)$ is hypercyclic, provided $|\lambda| \geq 1$; on the other hand, $T_{\lambda,b}$ is a convolution operator if and only if $\lambda = 1$. This leads us to the question: Is the set $\{T_{\lambda,b} \mid \lambda, b \in \mathbb{C}, |\lambda| \geq 1\}$ of hypercyclic operators closed under composition?

Third, it is known [3] that if there are no entire functions f such that f^k is hypercyclic for the translation operator τ_1, for any $k \geq 2$. On the other hand, the set of entire functions f such that f^k is hypercyclic for D, the differentiation operator, for every $k \in \mathbb{N}$ is a dense G_δ in $\mathcal{H}(\mathbb{C})$. So, let T be a hypercyclic operator on $\mathcal{H}(\mathbb{C})$. Is there some $f \in \mathcal{HC}(T)$ such that f^k is also in $\mathcal{HC}(T)$? Characterize those T for which such a phenomenon occurs.

5. Baire category and lineability

A recurring theme throughout our exposition has been the Baire category theorem. For instance, not only is there one continuous, nowhere differentiable function on $[0,1]$ but, by use of Baire's theorem, there is a dense G_δ of such functions. Not only is there one entire function f that is hypercyclic with respect to Birkhoff's translation operator but, by Baire's theorem, there is a dense G_δ of such functions. Not only is there one continuous function on the unit circle whose Fourier series at 1 diverges but, by Baire's theorem, there is a dense G_δ of such functions [33]. (F. Bayart [12] has shown that the set of such functions is lineable, and in fact D. Pérez, J. Seoane, and the author [9] have shown that the set is *algebrable*.)

As the reader can see, examples are plentiful. However, is this always true? And, if not, how can we tell when it is true that a residual set of vectors actually contains a 'big' vector space? We don't know.

We conclude our exposition by returning to Example 1 of Introduction 2. It is here that, for the first time in this survey, one encounters a discordant note. Indeed, V. Gurariy and L. Quarta recently proved the following result:

Theorem 14. [24] Consider the set $\mathcal{M} = \{f \in \mathcal{C}[0,1] \mid f$ attains its maximum at exactly one point of $[0,1]\}$. Then no subspace $V \subset \mathcal{M}$ can have dimension ≥ 2.

However, this is by no means a counterexample to any assertion related to the Baire category theorem since the set $\mathcal{M}$ is far from a G_δ. In fact, we are grateful to L. Quarta for pointing out to the following fact to us:

Proposition 15. $\mathcal{C}[0,1]\backslash\mathcal{M}$ is a dense G_δ.

Sketch of proof of Proposition 15. For each $n \in \mathbb{N}$, let

$$U_n = \{f \in \mathcal{C}[0,1] \mid \text{ for some } x \in [0,1],\ f(x) > \max_{|t-x|\geq 1/n} f(t)\}.$$

It is not difficult to see that U_n is open, for if $f \in U_n$ and $g \sim f$, then $g(x)$ will also be $> \max_{|t-x|\geq 1/n} g(t)$. Also, if $g \in \mathcal{C}[0,1]$ is arbitrary and $g(x_0) = \max_{t\in[0,1]} g(t)$, then by slightly increasing g near x_0 we get a function $h \sim g$, $h \in U_n$. In other words, each U_n is dense in $\mathcal{C}[0,1]$. Finally, $\mathcal{M} = \cap_n U_n$. Indeed, suppose that $f \in U_n$ for each n, but that f attains its maximum at two points, x_0 and x_1. For large n, we will have a contradiction since $f(x_0)$ cannot be strictly greater than $\max_{|t-x_0|>1/n} f(t)$. The converse inclusion is easy. Q.E.D.

Finally, we mention that Gurariy and Quarta show [24] that if we instead let $\mathcal{M} = \{f \in \mathcal{R} \mid f$ attains its maximum at precisely one point in $\mathbb{R}\}$, then it is known that $\mathcal{M}$ contains a 2–dimensional vector space. On the other hand, it is unknown if it is lineable, or even if it contains a 3–dimensional space.

References

1. Abakumov, E.; Gordon, J., *Common hypercyclic vectors for multiples of backward shift,* J. Funct. Anal. **200** (2003), no. 2, 494-504.
2. Ansari, S., *Existence of hypercyclic operators on topological vector spaces,* J. Funct. Anal. **148** (1997), no. 2, 384-390.
3. Aron, R.; Conejero, J. A.; Peris, A.; Seoane, J. B., *Powers of hypercyclic functions for some classical hypercyclic operators,* to appear.
4. Aron, R.; García, D.; Maestre, M., *Linearity in non-linear problems,* RACSAM Rev. R. Acad. Cienc. Exactas Fís. Nat. Ser. A Mat. **95** (2001), no. 1, 7-12.
5. Aron, R.; Gonzalo, R.; Zagorodnyuk, A., *Zeros of real polynomials,* Linear and Multilinear Algebra **48** (2000), no. 2, 107-115.
6. Aron, R.; Gurariy, V. I.; Seoane, J. B., *Lineability and spaceability of sets of functions on* $\mathbb{R}$, Proc. Amer. Math. Soc. **133** (2005), no. 3, 795-803.
7. Aron, R.; Hajek, P. *Zero sets of polynomials in several variables,* Archiv Math., **86** (2006), 561-568.
8. Aron, R.; Markose, D., *On universal functions.,* Satellite Conference on Infinite Dimensional Function Theory, J. Korean Math. Soc. **41** (2004), no. 1, 65-76.
9. Aron, R.; Pérez-García, D.; Seoane-Sepúlveda, J. B., *Algebrability of the set of non–convergent Fourier series,* Studia Math. **175** (1) (2006), 83-90.
10. Aron, R.; Rueda, M. P., *A problem concerning zero-subspaces of homogeneous polynomials,* Dedicated to Professor Vyacheslav Pavlovich Zahariuta. Linear Topol. Spaces Complex Anal. **3** (1997), 20-23.
11. Banakh, T.; Plichko, A.; Zagorodnyuk, A., *Zeros of quadratic functionals on non-separable spaces,* Colloq. Math. **100** (2004), no. 1, 141-147.
12. Bayart, F., *Topological and algebraic genericity of divergence and universal-*

ity, Studia Math. **167** (2005), 153-160.
13. Bernal González, L., *On hypercyclic operators on Banach spaces,* Proc. Amer. Math. Soc. **127** (1999), no. 4, 1003-1010.
14. Bernal González, L.; Montes-Rodríguez, A., *Non-finite-dimensional closed vector spaces of universal functions for composition operators,* J. Approx. Theory **82** (1995), no. 3, 375-391.
15. Birkhoff, G. D., *Démonstration d'un théorème élémentaire sur les fonctions entières,* C. R. Acad. Sci. Paris **189**, 473-475.
16. Bishop, E.; Phelps, R. R., *A proof that every Banach space is subreflexive,* Bull. Amer. Math. Soc. **67** 1961 97-98.
17. Bonet, J.; Peris, A., *Hypercyclic operators on non-normable Frchet spaces,* J. Funct. Anal. **159** (1998), no. 2, 587-595.
18. Bourdon, P., *Invariant manifolds of hypercyclic vectors,* Proc. Amer. Math. Soc. **118** (1993), no. 3, 845-847.
19. Fernández, G.; Hallack, A., *Remarks on a result about hypercyclic non-convolution operators,* J. Math. Anal. Appl. **309** (2005), no. 1, 52-55.
20. Fernández-Unzueta, M., *Zeroes of polynomials on* ℓ_∞, J. Math. Anal. Appl. **324**, 2 (2006), 1115-1124.
21. Fonf, V. P.; Gurariy, V. I.; Kadets, M. I., *An infinite dimensional subspace of* $C[0,1]$ *consisting of nowhere differentiable functions,* C. R. Acad. Bulgare Sci. **52** (1999), no. 11-12, 13-16.
22. Godefroy, G.; Shapiro, J., *Operators with dense, invariant, cyclic vector manifolds,* J. Funct. Anal. **98** (1991), no. 2, 229-269.
23. Costakis, G.; Sambarino, M., *Genericity of wild holomorphic functions and common hypercyclic vectors,* Adv. Math. **182** (2004), no. 2, 278-306.
24. Gurariy, V. I.; Quarta, L., *On lineability of sets of continuous functions,* J. Math. Anal. Appl. **294** (2004), no. 1, 62-72.
25. Gurariy, V. I.; Lusky, W., *Geometry of Mntz Spaces and Related Questions Series,* Lecture Notes in Mathematics, **1870**, 2005, XIII, 176 p.
26. Hallack, A., *Hypercyclicity for translations through Runge's Theorem,* to appear.
27. James, R. C., *Reflexivity and the supremum of linear functionals,* Ann. of Math. (2) **66** (1957), 159-169.
28. MacLane, G. R., *Sequences of derivatives and normal families. J. Analyse Math.,* **2**, (1952). 72-87.
29. Montes-Rodríguez, A., *Banach spaces of hypercyclic vectors,* Michigan Math. J. **43** (1996), no. 3, 419-436.
30. Plichko, A.; Zagorodnyuk, A., *On automatic continuity and three problems of The Scottish book concerning the boundedness of polynomial functionals,* J. Math. Anal. Appl. **220** (1998), no. 2, 477-494.
31. Rolewicz, S., *On orbits of elements,* Studia Math. **32** 1969 17-22.
32. Rudin, W., *Principles of Mathematical Analysis,* 3rd edition, McGraw-Hill (1976).
33. Rudin, W., *Real and Complex Analysis,* 3rd edition, McGraw-Hill (1987).
34. Soares, C. A., personal communication.

ALEXANDER GROTHENDIECK'S WORK ON FUNCTIONAL ANALYSIS

F. BOMBAL*

Departament of Mathematical Analysis
University Complutense
Madrid, Spain
E-mail: fernando_bombal@mat.ucm.es

Dedicated to the Memory of Miguel de Guzmán.

Alexander Grothendieck obtained the Medal Fields in 1966 for his contributions to Homological Algebra and Algebraic Geometry. However, Grothendieck's work on Functional Analysis, appeared in 25 papers between 1950 and 1957, had a tremendous and deep influence in the development of this area of Mathematics. Along this lecture, we shall try to give a perspective of Grothendieck's work on Functional Analysis, his ideas to study and introduce new properties in topological vector spaces and a quick look at part of his heritage.

1. Introduction

Alexander Grothendieck is one of the most influential mathematicians of the twentieth century. He received the **Field's** Medal in 1966 *"for his contributions to Homological Algebra and Algebraic Geometry"*, but this is to say little about the impact of Grothendieck's work in modern Mathematics. Quoting Ref. [3]:

> *The mere enumeration of Grothendieck's best known contributions is overwhelming: topological tensor products and nuclear spaces, sheaf cohomology as derived functors, schemes, K-theory and Grothendieck-Riemann-Roch, the emphasis on working relative to a base, defining and constructing geometric objects via the functors they are to represent, fibred categories and descent, stacks, Grothendieck topologies and topoi, derived categories, formalisms of*

*Work partially supported by Spanish MCyT grant MTM2004-01308.

local and global duality, étale cohomology and the cohomological interpretation of L-functions, crystalline cohomology, "standard conjectures", motives and the "yoga of weights", tensor categories and motivic Galois groups. It is difficult to imagine that they all aprang from a single mind.

The above list is impressive, and surely part of the theories mentioned are unfamiliar to most readers. It is even more impressive to know that the list of "official" Grothendieck publications begins in 1950 (with a paper that appeared in *C. R. Acad. Sci. Paris*), and ends in 1974 (*Groupes de Barsottti-Tate et cristaux de Dieudonné*, Séminaire de Mathématiques Supérieures. **45** (Été 1970). Les Presses de l'Université de Montréal, 1974).[a]

Several colleagues and friends refer to his total dedication to research, living alone and working for 25 or 26 hours each "day". Hence, why did he abruptly end a career so fertile at the age of 42? The official reason given by Grothendieck himself for resigning his position at the *Intitut des Hautes Etudes Scientifiques (IHES)* was that he had discovered that the Ministry of Defense had partly subsidized the Institute. But it seems that this reaction is, at least, exaggerated. **P. Cartier**, a friend and colleague at the IHES, refers in Ref. [4] his opinion:

> *...He* ([Grothendieck]) *is the son of a militant anarchist who had devoted his life to revolution... He lived as an outcast throughout his entire childhood and was a "displaced person" for many years, traveling with a United Nations passport* ([his citizenship papers disappeared in Berlin, during 1945]... *He lived his principles, and his home was always wide open to "stray cats". In the end, he came to consider Bures-sur-Yvette* [where was located the IHES] *a gilded cage that kept him away from real life. To this reason, he added a failure of nerve, a doubt as to the value of scientific activity... He confided his doubts to me and told me that he was considering activities other than mathematics. One should perhaps add the effect of a well-known "Nobel syndrome"... yielding to the pernicious view that sets 40 as the age when mathematical creativity ceases. He may have believed that he had passed his peak and that thenceforth he would be able only to repeat himself with less*

[a]A list of all Grothendieck's papers on Functional Analysis is included as Appendix at the end of the paper

effectiveness.

The mood of the time also had a strong influence. The disaster that had been the second Viet Nam war, from 1963 to 1972, had awakened many consciences... A significant number of French mathematicians took concrete action and traveled to Hanoi, as he (and I) did... The cold war was at its height, and the risk of a nuclear confrontation was very real. The problems of overpopulation, pollution, and uncontrolled development -everything that is now classified as ecology- had also begun to attract attention. There were plenty of reasons to call science into question!

I think that Cartier's remark shed some light on Grothendieck's character and his personal circumstances. For the sake of completeness, it would probably be convenient to take a quick look at Grothendieck's timeline. But before proceeding, one remark: Most of the available biographical information about Grothendieck coincides with the main facts, but there is a constant reference to the inaccuracies of the *other* biographies. I have followed essentially the biographical notes included in Ref. [4] and the information that appears on the web page `http://www.math.jussieu.fr/~leila/biog.html` entitled *a Brief Timeline for the Life of Alexander Grothendieck* (with the added subtitle: *which has the advantage of accuracy*).

Alexander Grothendieck was born in Berlin, in 1928. His father, **Alexander Schapiro** (called Sascha) was a Jewish revolutionary born in Russia who participated in many of the revolutionary movements that occurred in central Europe during the first decades of the twentieth century. In the 1920's he lived in Germany, working as a street photographer and fighting politically against Hitler and the Nazis. There he met **Johanna Grothendieck** (called Hanka), a German Jewish woman sharing Sascha's ideals. After 1933, the couple fled to Paris, leaving young Alexander with a foster family in Hamburg. Sascha and Hanka joined the anarchists of the F.A.I. when the Spanish Civil War broke out in 1936. They returned to France in 1939 and Hanka found work in Nîmes.

Due to the political situation Alexander's foster mother sent him to join his parents in France in 1939. After the French defeat in 1940, Vichy's collaborationist government promulgated anti-Jewish laws for the unoccupied zone, and Hanka and her son Alexander were interned in the Rieucros Camp. Sascha was interned in Le Vernet camp and then sent directly to Auschwitz, where he died in 1942.

Also in 1942, the Rieucros camp was dissolved. Alexander was sent to the village Le Chambon sur Lignon and was housed at the Swiss Foyer, attending the *Collège Cévenol.* He studied mathematics at the University of Montpellier from 1945 to 1948 and, having completed his bachelor's degree, went to Paris for his doctorate, with a letter of recommendation to **Élie Cartan**. His son, **Henri Cartan** advises Grothendieck to do his doctorate in Nancy with **Laurent Schwartz**. He finished his Thesis (a real masterpiece) in 1953 and spent the following two years in Brazil. In 1955 he visited the Universities of Kansas and Chicago and hoped to find a position in France, in spite of his foreign nationality.

In 1959, the IHES was created in Bures-sur-Yvette and a research position was offered to Grothendieck. During the 12 years spent there, he completely renewed Algebraic Geometry. His *Élements de Géométrie Algébrique* (he wrote 4 volumes with Dieudonné) and the series *Séminaire de Géométrie Algébrique* (one per year, from 1960-61 to 1967-68; some of them divided in several volumes) form an epoch-making contribution.

As we have said, Grothendieck left the IHES in 1970. After that, he became vividly interested in Ecology for some time, and founded the group *Survivre et Vivre.* For two years, he had a temporary post at the prestigious *Collége de France* in Paris, but he used his lectures to talk more about questions of ecology and peace than of mathematics.

In 1972 Grothendieck obtained French citizenship and lectured at Montpellier University from 1973 to 1984. In 1984 he applied for a position of Director or Research at CNRS, but specified that he did not want any regular research duty. After some long discussion at the National Committee, he was given the position.

In 1988 he officially retired. In the same year he was awarded the Crafoord Prize (jointly with his student **Pierre Deligne**) from the Swedish Royal Academy of Sciences, but he declined the prize on ethical grounds. A letter explaining his reasons, appeared in 1989 (see Ref. [14]).

In 1991 he left his home suddenly and disappeared. He is said to live in some part of the Pyrenées, refusing virtually all human contact and spending his time dedicated to meditation on philosophical questions. The monumental manuscript *Récoltes et Semailles* ("Harvesting and Sowing"), a very personal kind of autobiography of more than 2000 pages, is now available on the Internet in several languages.

A large amount of mathematical and biographical information with photos and links to related sites may be found at `http://www.grothendieck-circle.org/`.

2. Grothendieck and Functional Analysis

Grothendieck's contributions to Algebraic Geometry are well known by general mathematicians. A great part of the three volumes in Ref. [3], published on the occasion of Grothendieck's sixtieth birthday, are devoted to these aspects of his work. Ref. [1] is also a good survey (in Spanish) on Grothendieck's work, focusing mainly on those topics related to Algebraic Geometry. In the following pages, I shall try to recount some contributions from the beginning of his career: those related to Functional Analysis. They are included in 24 papers and one book, that appeared between 1950 and 1957 (except a short note about the trace of certain operators between Banach spaces, published in 1961). Among them are some of the most influential works in the development of Functional Analysis.

I shall focus mainly on his contributions to the theory of topological tensor products, first in the setting of locally convex spaces (the subject of his Thesis) and later as a powerful tool to study the structure of Banach spaces (mainly contained in the famous Sao Paolo's *Résumé*). Between these two masterpieces, I shall briefly comment on the remarkable paper on $\mathcal{C}(K)$-spaces which appeared in the *Canadian J. Math.*, where Grothendieck axiomatizes some important properties. I will refer to this not only for its transcendence in the future development of the theory, but because I think it is a paradigmatic example of Grothendieck's way of thinking. In any case, I shall try not to be overly technical, and will always give priority to clarity over precision.

2.1. *The Thesis: "Produits tensoriels topologiques et espaces nucléaires."*

As we have seen, Grothendieck went to Paris to do his doctorate in 1949. After attending some courses there, he followed the advise of H. Cartan and went to Nancy to work with **Laurent Schwartz** and **Jean Dieudonné**. Allow Schwartz himself describe his impressions:

> *We* [Dieudonné and Schwartz] *received Grothendieck in october, 1951. He showed a 50 pages paper on "Integration with values in a topological group" to Dieudonné. It was correct, but uninteresting. Dieudonné, with all the aggressiveness he could have,* ([and he could a lot] *gave him a severe tongue-lashing, arguing that he should not work in such a manner, just generalizing for the pleasure of doing so...Dieudonné was right, but Grothendieck never admitted it...*

> *We had just published a paper on "Les espaces $\mathcal{F}$ et $\mathcal{LF}$"... that included 14 questions, problems that we were unable to solve. Dieudonné proposed to Grothendieck that he think about some of them, those that he preferred. We did not see him for some weeks. When he appeared again, he had solved half of the questions! Deep and difficult solutions which needed new notions. We were amazed.* (Ref. [16, p. 292-293].)

Schwartz realized at once that he had met a mathematician of first order and in the spring of 1953 he proposed to Grothendieck, as a subject of his Thesis, the general problem of finding a "good" topology on the tensor product $E \bigotimes F$ of two locally convex spaces E and F. At this time, Schwartz was starting the theory of vector-valued distributions, that is, the study of the space $\mathcal{D}'(F) := \mathcal{L}(\mathcal{D}, F)$ of continuous linear operators from the test space $\mathcal{D} = \mathcal{D}(\mathbb{R}^n)$ (the space of C^∞ scalar functions with compact support, endowed with its usual inductive limit topology) into the locally convex space F. A good topology for this space was evident, inducing a corresponding good topology on its dense subspace $\mathcal{D}' \otimes F$. But this was not obvious in general.

Grothendieck spent the summer of 1953 in Brazil and at the end of July he wrote Schwartz a, in certain sense, deceptive letter: on $E \otimes F$ there were two topologies, one as natural as the other, yet different! Schwartz did not know what to say, because on $\mathcal{D}' \otimes F$ there was *one* natural topology. Fortunately, two weeks latter a triumphal letter arrived: the two natural topologies coincide on $\mathcal{D}' \otimes F$ (and with the natural one)!

The two natural topologies discovered by Grothendieck are what we now know as the *projective* or π-topology and the *injective* or ϵ-topology. The π-topology is the greatest "reasonable" one (in some precise sense), and makes, with the natural identifications, $\mathcal{L}(E \otimes_\pi F, G) = \mathcal{B}(E \times F, G)$, all the continuous bilinear maps from $E \times F$ into G. The ϵ-topology is the least "reasonable" topology and has a great importance in the development of important classes of operators.

Grothendieck's Thesis was completed in 1953. It is a masterpiece of more than 300 pages which contains not only the main theorems of the theory of topological tensor products, but also new methods, techniques and a lot of seminal ideas which were to renew Functional Analysis. But let us hear the opinion of his Thesis' advisor, L. Schwartz:

> ... *"It is a monument, a masterpiece of the first order. It was necessary to read it, to understand it, to learn from it, because it*

was difficult and deep. It took to me six months at full time. What a hard job, but what a joy!... I learned a lot of new things. It was the most beautiful of "my" *Thesis...*" (Ref. [16, p. 294].)

Besides the definition of the π and ϵ topologies on a tensor product of locally convex spaces and a deep study of these new objects, examples and applications to the study of vector-valued function spaces, etc., the Thesis contains much more. We shall mention some of the contents:

2.1.1. *I.- The approximation property.*

Since the π-topology is finer than the ϵ-topology, we can extend the identity operator to the respective completions, obtaining a canonical map

$$E\widehat{\otimes}_{\pi}F \to E\widehat{\otimes}_{\epsilon}F,$$

not necessarily injective, as Grothendieck points out. He says that he does not know any example where the injectivity fails, and poses the *Problème de biunivocité* and its relative, the *Problème d'approximation.* He gives a great number of different formulations of these problems and introduces a sufficient condition to answer both questions affirmatively: E or F to have the *condition d'approximation* (approximation property), which means that the identity operator can be uniformly approximated on precompact subsets by operators of finite rank. When considering Banach spaces, he introduces also the *metric approximation property* (requiring that the approximating operators have norm ≤ 1). He proves that the classical Banach spaces, their duals, biduals, etc. have the metric approximation property. The nuclear spaces and most of the usual locally spaces that appear in the Theory of Distributions also have the approximation property. Then Grothendieck states as an important problem to know if *every* locally convex space (equivalently, every Banach space) has the approximation property. He gives several equivalent conditions and different properties of permanence, showing in particular that it is equivalent to the problem No. 153 of the famous *Scottish Book*, in which **S. Banach** in 1935 collected the problems posed by the group around Banach and Steinhaus in Lwow and their invited visitors. Problem 153 was posed in November 6, 1936 by **S. Mazur**, the prize offered for the solution being a live goose.In fact, Mazur had the opportunity of giving the prize in 1973 to **Per Enflo**, who had solved the conjecture in the negative the year before. Grothendieck's work was fundamental in later work and its negative solution.

2.1.2. *II.- Nuclear, integral and related operators.*

When E, F are Banach spaces, the natural inclusion $E' \bigotimes F \hookrightarrow \mathcal{L}(E, F)$ can be extended (being $\mathcal{L}(E, F)$ complete with its usual norm) to a map $E'\widehat{\otimes}_\pi F \to \mathcal{L}(E, F)$. The operators in the image of this map are called by Grothendieck *nuclear operators.* In the general case, an operator $T : E \to F$ between locally convex spaces is called *nuclear* if it can be factorized in the form $T = A \circ S \circ B$, with S a nuclear operator between two Banach spaces. Grothendieck made a deep study of this class of operators, and gave many examples.

Another important class of operators isolated by Grothendieck is that of *integral operators:* Since the ϵ-topology is coarser that the π one, the topological dual $(E\widehat{\otimes}_\epsilon F)'$ is a subset $J(E, F) \subset \mathcal{B}(E \times E, F)(\equiv (E\widehat{\otimes}_\pi F)')$. The bilinear forms in $J(E, F)$ are called *integral* by Grothendieck, who also gives a characterization in terms of an integral representation. A linear operator $T : E \to F$ is *integral* if the corresponding bilinear form on $E \times F'$ is integral. In case of Banach spaces, Grothendieck immediately gives a factorization criterium: $T : E \to F$ is integral if and only if the composition of T with the canonical embedding of F in its bidual F'' can be factorized in the form

$$E \to L_\infty(\mu) \xrightarrow{i} L_1(\mu) \to F'',$$

where μ is a probability on some compact space and i is the natural inclusion.

This method of factoring an operator through classical Banach spaces in order to take advantage of the knowledge of these spaces was probably not due to Grothendieck, but he made a systematic use of it, and this is part of his legacy to Functional Analysis.

Grothendieck carries out a deep study of integral maps, with surprising applications to classical analysis, summable sequences, vector measures, etc.

The nuclear maps are always compact and integral, and the composition of two integral maps is nuclear.

In the case of Banach spaces Grothendieck also introduced other interesting classes of operators: the *right* (resp., *left*) *semi-integral* (or "pre-integral") operators $T : E \to F$, just imposing that the composition with an embedding of F into a L_∞-space (resp., with a quotient map from an L_1-space onto E) be integral. The right semi-integral operators are precisely the familiar *absolutely summing operators*, generalized and studied in the 1960's by **Mityagin, Pelczynski** and **Pietsch**, among others, to the important class of *absolutely p-summing operators* ($1 \leq p < \infty$).

2.1.3. *III. The kernel theorem and nuclear spaces.*

During the International Congress of Mathematics of 1950, Schwartz had announced his surprising *Théorème des noyaux* ("Kernel Theorem"), asserting that *every* continuous linear operator $T \in \mathcal{L}(\mathcal{D}, \mathcal{D}')$ ($\equiv \mathcal{B}(\mathcal{D} \times \mathcal{D})$, i.e., the continuous bilinear maps on $\mathcal{D} \times \mathcal{D}$), came from a "distributional kernel", that is, a distribution in two variables $K(x,y) \in \mathcal{D}'(\mathbb{R}^n \times \mathbb{R}^m)$ such that for $\varphi \in \mathcal{D}(\mathbb{R}^n)$ and $\psi \in \mathcal{D}(\mathbb{R}^m)$,

$$T(\varphi)(\psi) = \langle \varphi(x)\psi(y), K(x,y) \rangle = \langle \varphi \otimes \psi, K \rangle$$

that could be written formally as $\int K(x,y)\varphi(x)\psi(y)\, dxdy$. This is rather surprising as it is not true for most of the usual function spaces. For instance, the identity operator in the usual L_2 space cannot be expressed as a kernel operator.

In general, since the pioneering works of **D. Hilbert** and **F. Riesz** it was well known that the operators on some function space, like L_2, of the type

$$T(f)(x) := \int K(x,y) f(y)\, dy$$

(*kernel operator*) had especially good properties, but unfortunately they did not exhaust all the possible operators. And by the way, this was one of the main difficulties in the rigorous formulation of Quantum Mechanics. The idea of **P. Dirac** and others was to express any *observable* ($\sim$ "linear operator") in terms of a base formed by the "states" (ψ_p) of the system (the eigenfunctions corresponding to the eigenvalues of the operator, that is, the "spectrum" of the observable). But in most cases this spectrum was not countable (in words of Dirac: *"... the total number of independent states is infinite, and equal to the number of points of a line"*). Then the corresponding "matrix" (α_{pq}) was indexed by two continuous parameters, $p, q \in \mathbb{R}$, and the action of the observable α on some state ψ_q was written in the form

$$\alpha(\psi_q) = \int_{\mathbb{R}} \alpha_{pq} \psi_p\, dp,$$

that is, as a kernel operator. But then, if one wants to represent in this way the "multiplication by a non zero constant c" operator, we need a kernel of the type $\alpha_{pq} = c\delta(p-q)$, that is, a *singular function*. Dirac used also other singular functions and its derivatives, just applying formally the method of "integration by parts".

Hilbert tried to follow a similar method, but the appearance of singular functions, made him to look for another point of view. This was the **J. von Neumann** spectral theory of (not necessarily bounded) operators on subspaces of a Hilbert space.

The kernel theorem allows us to justify rigorously part of Dirac's ideas: in fact, if E and F are function spaces over open subsets $U \subset \mathbb{R}^n$, $V \subset \mathbb{R}^m$ usually we have

$$\mathcal{D}(U) \hookrightarrow E \hookrightarrow \mathcal{D}'(U)$$

and

$$\mathcal{D}(V) \hookrightarrow F \hookrightarrow \mathcal{D}'(V)$$

(with continuous embeddings, the first one with dense range). Hence, every continuous operator $S : E \to F$ gives rise by composition to a continuous operator $T : \mathcal{D}(U) \to \mathcal{D}'(V)$. Therefore, it can be represented by a (distributional) kernel operator!

This long detour tries to explain why a great part of Grothendieck's thesis is devoted to the study and characterization of those locally convex spaces E such that $E \otimes_\pi F = E \otimes_\epsilon F$ for every locally convex space F. He called such spaces *nuclear spaces.* The reason is that the spaces $\mathcal{D}$ (and $\mathcal{D}'$) are nuclear, and the kernel theorem is a trivial consequence of this fact and the (easy) result that $\mathcal{D}(U) \otimes_\epsilon \mathcal{D}(V)$ is a *dense* topological subspace of $\mathcal{D}(U \times V)$: In fact, let $T : \mathcal{D}(U) \to \mathcal{D}'(V)$ be a continuous linear map, and let $B : \mathcal{D}(U) \times \mathcal{D}(V) \to \mathbb{K}$ be the corresponding continuous bilinear map. Then,

$$B \in \mathcal{B}(\mathcal{D}(U)\times\mathcal{D}(V),\mathbb{K}) = (\mathcal{D}(U)\widehat{\otimes}_\pi\mathcal{D}(V))' = (\mathcal{D}(U)\widehat{\otimes}_\epsilon\mathcal{D}(V))' = \mathcal{D}'(U\times V)$$

(where the $\widehat{}$ means the completion of the corresponding space), which is essentially the content of the kernel theorem *via* the natural identifications.

Grothendieck carries out a deep study of this class of locally convex spaces (which contains no Banach space of infinite dimension), proving that it enjoys very good stability and permanence properties, giving many examples and applications.

By the way, this is a typical example of Grothendieck's way of doing mathematics: to put the problem in a more general setting and find a general theory (usually, very deep and far-reaching), which contains the solution of the initial problem as a particular case. Of course, this is the way most of twentieth century mathematics was developed, but usually the general theories were created by many authors along a certain time. In Grothendieck's work, this is a constant procedure!

Grothendieck's Thesis [8] contains, of course, much more. I have simply tried to take a quick look at it, mentioning some of the most relevant results. It also contains a great number of open questions and problems that motivated a great deal of research activity as Grothendieck's work became be known among specialists. The Thesis appeared in 1955, published as Vol. No. 16 of the prestigious *Memoir of the American Mathematical Society.* Since it took such a long time to be published, Grothendieck wrote a survey (Ref. [9]) quoting some of the more relevant results and as available references for his later works on the subject.

2.2. *The Dunford-Pettis and relatives properties.*

In 1953 [10] appeared in the Canadian Journal of Mathematics a paper "*...devoted essentially to the study of the weakly compact linear operators from a $\mathcal{C}(K)$-space into an arbitrary locally convex space F.* (Ref. [10, introduction]). By transposition, this is equivalent to the study of weakly compact subsets of the space of Radon measures on K (the dual of $\mathcal{C}(K)$). In fact, the paper contains some of the most useful weak compactness criteria on the spaces of Radon measures. But it contains much more.

Grothendieck's favorite method for studying general classes of operators by factoring them through "classical" spaces of type $\mathcal{C}(K)$, $L_1(\mu)$ or Hilbert spaces (used in his *Thesis*, but much more in the *Résumé*), made it quite important to know the behavior of different classes of operators on these spaces. There are several important results in this direction in the paper, starting with a Riesz-type representation in terms of vector measures. He remarked that, from Riesz's classical representation theorem, any operator $T : C(K) \to E$ could be represented in the form

$$T(f) = \int_K f\, dm$$

where m is a regular, finitely additive vector measure with finite semi-variation on the Borel subsets of K, with values in E'' (the *representing measure* of T). Grothendieck proved that T is weakly compact if and only if its representing measure takes values in E or, equivalently, it is countably additive.

The relationships between properties of the operator and its representing vector measure are widely used in later work on this subject, being very fruitful for both theories: linear operators on $\mathcal{C}(K)$ spaces (and also on vector valued function spaces $\mathcal{C}(K, E)$) and vector measures.

On the other hand, the article emphasizes the "functorial" point of view of Grothendieck: In order to study the structure of some mathematical object, you have to look at the behavior of the morphisms on and into it. This was quite usual in some parts of Mathematics, but not so in Analysis. The paper contains the first systematic treatment of what I called in [Bo] the *homological method* for defining properties on Banach spaces (Grothendieck treats the general case of operators between locally convex spaces, but he also mentions that *"Ce travail pu se traiter sans sortir du cadre des espaces de Banach."*).

The general scheme, as exposed in [Bo], is the following: Let Θ, Φ be two classes of linear operators between Banach spaces (in such a way that $\Theta(E, F)$, $\Phi(E, F)$ denote subsets of $\mathcal{L}(E, F)$ for every pair E, F of Banach spaces), and let $\mathcal{E}$ be a certain class of Banach spaces. We shall say that E *has property* $P(\Theta, \Phi; \mathcal{E})$, and we shall write $E \in P(\Theta, \Phi; \mathcal{E})$, if

$$\Theta(E, F) \subset \Phi(E, F), \quad \text{for every } F \in \mathcal{E}.$$

(When $\mathcal{E}$ is the class of *all* Banach spaces we shall omit its mention, writing simply $P(\Theta, \Phi)$). Clearly, the property considered could be interested only when the defining relation does not hold trivially.

Usually, the classes Θ and Φ have some structure; more concretely, they are usually *operator ideals* (which means that the class is stable under composition with continuous linear operators and $\Theta(E, F)$ is a vector subspace of $\mathcal{L}(E, F)$, containing the finite rank operators). In this case, property $P(\Theta, \Phi; \mathcal{E})$ is an isomorphic invariant, stable under finite products and passing to complemented subspaces. It is easy to prove that when Φ is a *surjective* operator ideal, $P(\Theta, \Phi; \mathcal{E})$ is also stable under the formation of quotients by closed subspaces.

In order to give some examples, let us consider the following classes of operators:

-$\mathcal{L}$: all the operators.

-$\mathcal{K}$: the *compact* operators, i.e., those sending bounded set into relatively compact subsets.

-$\mathcal{W}$: the *weakly compact* operators, i.e., those sending bounded sets into *weakly* relatively compact subset.)

-$\mathcal{DP}$: the *Dunford-Pettis* (or *completely continuous*) operators, i.e., those that send weakly convergent sequences into norm convergent ones (equivalently, they transform weakly compact subsets into norm compact subsets.)

-$\mathcal{D}$: the *Dieudonné* (or *weakly completely continuous*) operators, i.e., those which transform weakly Cauchy sequences into weakly convergent ones.

All the above classes are operators ideals, closed under the usual operator norm. Besides, $\mathcal{K}$ and $\mathcal{W}$ are surjective. It is also clear that

$$\mathcal{K} \subset \mathcal{W} \subset \mathcal{D} \subset \mathcal{L} \tag{1}$$

and

$$\mathcal{K} \subset \mathcal{DP} \subset \mathcal{D} \subset \mathcal{L}, \tag{2}$$

with strict inclusions, and with no other general relation. With our notations, we obviously have:

- $E \in P(\mathcal{L}, \mathcal{K})$ if and only if E is finite dimensional.

-$E \in P(\mathcal{L}, \mathcal{W})$ if and only if E is reflexive.

-$E \in P(\mathcal{L}, \mathcal{DP})$ if and only if weakly convergent sequences in E are norm convergent, i.e., E is a *Schur* space.

-$E \in P(\mathcal{L}, \mathcal{D})$ if and only if E is weakly sequentially complete.

2.2.1. *The Dunford-Pettis property.*

This was the first property introduced by Grothendieck in the paper we are considering. His motivation was a long 1940's article by **N. Dunford** and **J. Pettis** in which they proved that weakly compact operators on L_1-spaces were (in our notation) Dunford-Pettis operators (i.e., $L_1 \in P(\mathcal{W}, \mathcal{DP})$). Important consequences followed from this fact. Grothendieck axiomatized the property and called it the *Dunford-Pettis Property* (DPP for short). He gave several equivalent formulations and proved immediately that the property passes from E' to E. Since the dual of an L_1 space (built over a Radon measure) is an L_∞ space, hence isomorphic to a $\mathcal{C}(K)$ space, it is enough to prove that this last space enjoys the DPP for recovering the Dunford and Pettis' result. This is one of the important results contained in Grothendieck's memoir.

The DPP is "far" from reflexivity, since reflexive spaces with the DPP are finite dimensional. Also, the DPP can be localized, in the sense that it coincides with the property $P(\mathcal{W}, \mathcal{DP}; \{c_0\})$.

The DPP has been extensively studied (we refer the interested reader to the survey Ref. [6]), and it gives important information on the structure of the spaces having it . Besides the $\mathcal{C}(K)$ and L_1-spaces, the disc algebra A

(the space of all continuous functions on the unit disc $D := \{z \in \mathbb{C} : |z| \leq 1\}$, which are analytic in the open unit disc), their analogous d-dimensional, the ball algebra $A(B_d)$ and the polydisc algebra $A(D^d)$, and the space H^∞ of all bounded analytic functions on the open unit disc, enjoy the DPP. The last three results are due to **J. Bourgain**, in a series of deep papers published in 1983-84 in *Studia Math.* and in *Acta Math.* He also gave a correct proof of a result announced by Grothendieck: the space $C^k(U)$ of all complex-valued functions which are continuous with all their derivatives or order $\leq k$, on a d-dimensional compact manifold U, has the DPP.

2.2.2. *The reciprocal Dunford Pettis and the Dieudonné properties.*

Following the same idea, Grothendieck introduces two more properties. In our notation:

-$P(\mathcal{DP}, \mathcal{W})$: the *reciprocal Dunford-Pettis property* (RDPP for short.)

-$P(\mathcal{D}, \mathcal{W})$: the *Dieudonné property* (DP for short; the name is due to a Dieudonné's result on weakly convergent sequences of Radon measures).

In reality, any of the above properties are equivalent by duality to a weak compactness criteria in the dual of the space enjoying it. And this is the way that Grothendieck proved that $\mathcal{C}(K)$-spaces enjoy both properties. He also proved that both properties were stable under complemented subspaces, finite products and quotients (a trivial consequence of the mentioned fact that $\mathcal{W}$ is a surjective operator ideal).

¿From the inclusions (1) and (2) it is clear that DP implies the RDPP. Also, a weakly sequentially complete space E (that is, $E \in P(\mathcal{L}, \mathcal{D})$) has the DP if and only if $\mathcal{L}(E, \cdot) = \mathcal{W}(E, \cdot)$, i.e., E is reflexive. Consequently, no infinite dimensional L_1-space enjoys the DP. And, since such a space contains a complemented copy of ℓ_1, it also fails the RDPP. On the other hand, Rosenthal's dichotomy theorem yields immediately that if E contains no copy of ℓ_1, it enjoys the DP and the RDPP.

Grothendieck gave several applications of his results to the structure of classical Banach spaces.

2.2.3. *The "Grothendieck spaces" and the hereditary properties.*

The last chapter in Grothendieck's paper is devoted to some particular classes of $\mathcal{C}(K)$ spaces. In the first part he considers the case when K is a *stonean space* (or *extremally disconnected*), what means that the closure of every open set is open. This is equivalent to the fact that $\mathcal{C}_{\mathbb{R}}(K)$ is a complete lattice for the usual pointwise order. A typical example is the Stone-Cech compactification of any discrete topological space. Every $L_\infty(\mu)$-space is isomorphic to a $\mathcal{C}(K)$-space with K stonean.

Grothendieck proved that *any* continuous linear operator from such a $\mathcal{C}(K)$-space into a *separable* Banach space, is weakly compact. In other words, if $\mathcal{S}$ denotes the class of all separable Banach spaces, the spaces $\mathcal{C}(K)$ with K stonean verify property $P(\mathcal{L}, \mathcal{W}; \mathcal{S})$. These spaces are now known as *Grothendieck spaces.* An internal characterization (obtained also by Grothendieck) is that weak* convergent sequences in the dual space are weakly convergent. Obviously, reflexive spaces are Grothendieck (and they are the only separable Grothendieck spaces). Let us add that **J. Bourgain** proved in 1983 that H^∞ is a Grothendieck space, and that the property can also be localized: it coincides with the $P(\mathcal{L}, \mathcal{W}; \{c_0\})$.

The last important result included in the paper asserts that every subspace of c_0 enjoys the DPP, the DP and the RDPP. This is the prototype of the so-called (for obvious reasons) "hereditary properties". A crucial lemma for the proof is that every normalized weakly null sequence in c_0 contains a basic sequence equivalent to the usual c_0-basis.

The seminal ideas contained in this paper were not well appreciated by Banach space researchers for more than 10 years, but then they became tremendously influential in the development of the theory.

As for the "homological method", let us mention that $E \in P(\mathcal{W}, \mathcal{K})$ is equivalent to E' being Schur and, by a result of Odell and Rosenthal, $E \in P(\mathcal{DP}, \mathcal{K})$ if and only if contains no copy of ℓ_1. Of course, when considering new classes of operators one obtains new properties (see Ref. [2].)

2.3. *The Sao Paulo's "Résumé"*

Surely many specialists in Banach spaces will subscribe the opinion of **A. Pietsch**, that this is *the most spectacular paper of modern Banach space theory* (and one of the most influential, I would add). It was submitted to the Bulletin of the Sao Paolo's Mathematical Society in June 1954, and it appeared in 1956, but was not grasped for more than 10 years. In 1968 a long paper of more than 50 pages [15] appeared in *Studia Mathematica* intending to show to the mathematical community some of the jewels hidden in the *Résumé.* The authors wrote in the introduction:

> *"The main purpose of the present paper is to give a new presentation as well as new applications of the results contained in Grothendieck's paper...*

Though the theory of tensor products constructed in Grothendieck's paper has its intrinsic beauty we feel that the results of Grothendieck and their corollaries can be more clearly presented without the use of tensor products....

The paper of Grothendieck is quite hard to read and its results are not generally known even to experts in Banach space theory..."

In fact, the authors bypassed the language of tensor products by systematically using what is now konwn as *p-summing operators*, whose foundation had appeared in another seminal paper of **A. Pietsch** published also in *Studia* in 1967.

But let us come back to Grothendieck's *Résumé*. The underlying idea in the paper is to obtain new classes of operators between Banach spaces by defining suitable norms on $E \otimes F$. When E and F are finite-dimensional, $E' \otimes F = \mathcal{L}(E, F)$, and a norm in $E' \otimes F$ defines an operator norm. The extension of this procedure to infinite-dimensional spaces is not trivial. It involves a skillful use of the so called *trace duality* (in the finite dimensional case this essentially means the duality between the spaces $E \otimes F \equiv L(F^*, E)$ and $(E \otimes F)^* \equiv L(E, F^*)$, given by the trace of the composition). Obviously, this cannot be trivially extended to the setting of infinite-dimensional Banach spaces and continuous linear maps. Grothendieck was aware of this problem and he devoted the first chapter of the Memoir to establish a method for defining "good" norms on a tensor product of Banach spaces: the $\otimes$-norms (or *tensor norms*). Such a norm $\|\cdot\|_\alpha$ should be in the firs place, *reasonable*, which means that $\|x \otimes y\|_\alpha = \|x\|\|y\|$ and $x' \otimes y' \in (E \otimes_\alpha F)'$, with (dual) norm $\|x' \otimes y'\|_{\alpha'} = \|x'\| \; \|y'\|$ (hence, $E' \otimes F'$ is a subspace of $(E \otimes_\alpha F)'$. The dual norm α' induces then on $E' \otimes F'$ another reasonable norm.)

Of course, the norms ϵ and π are reasonable (and duals one of the other). In fact, a norm α defined for every pair of normed spaces is reasonable if and only if $\epsilon \leq \alpha \leq \pi$.

On the other hand, the $\otimes$-norms should verify a good functorial property: the so called *metric mapping property*. This means that whenever $u_i \in \mathcal{L}(E_i, F_i)$ $(i = 1, 2)$, then $u_1 \otimes u_2 \in \mathcal{L}(E_1 \otimes_\alpha E_2, F_1 \otimes_\alpha F_2)$, with norm $\leq \|u_1\| \; \|u_2\|$.

Next, Grothendieck gives a method to construct $\otimes$-norms with good duality properties: First, he considers a $\otimes$-norm α defined on the class FIN of all the *finite-dimensional* Banach spaces. Then, he *extends* this norm to every pair of Banach spaces in the following way: If E and F are normed spaces, for $u \in E \otimes F$ we define

$$\|u\|_{\overrightarrow{\alpha}} := \inf\{\|u\|_\alpha : u \in M \otimes N\},$$

when M and N run over the finite dimensional subspaces of E and F, respectively. This is what now is known as the *finite hull procedure* for extending a tensor norm from the class FIN to the class NORM or all normed spaces. There is another standard procedure (the *cofinite hull*, essentially due to **H. P. Lotz**) to extend a tensor norm α on FIN to a tensor norm $\overleftarrow{\alpha}$ on NORM (see Ref. [5, Ch. II]), that was not considered by Grothendieck. And there is a good reason for that. In fact, if E and F have the approximation property, $E \otimes_{\overrightarrow{\alpha}} F = E \otimes_{\overleftarrow{\alpha}} F$, and it was 20 years after Grothendieck's *Résumé* that the first space without the approximation property was discovered!

Grothendieck considers some operations with the $\otimes$-norms on FIN, which are extended to NORM by the finite hull procedure. In particular, when $M, N \in$ FIN, for every $\otimes$-norm α, $M \otimes N = (M' \otimes_\alpha N')'$ (algebraically). The dual norm induced on $M \otimes N$ by this identification is denoted by α' and called the *dual norm* of α. It is also a $\otimes$-norm and its extension $\overrightarrow{\alpha'}$ is called the dual tensor norm of $\overrightarrow{\alpha}$. Since $\alpha'' = \alpha$ on FIN, the same relation holds for their extensions.

Now Grothendieck proceeds to define the class of operators and bilinear forms associated to a tensor norm α: since $\alpha \leq \pi$, for every pair of Banach spaces E, F, the dual of the completion $E\widehat{\otimes}_\alpha F$ of $E \otimes_\alpha F$ can be identified to a subspace of $\mathcal{B}(E, F)$, the dual of $E\widehat{\otimes}_\pi F$. Grothendieck calls a bilinear form B on $E \times F$ of *type* α if it belongs to the dual of $E\widehat{\otimes}_{\alpha'} F$. Its norm in this dual is denoted by $\|B\|_\alpha$. Analogously, a linear map $u : E \to F$ is of *type* α if its canonically associated bilinear map on $E \times F'$ is of type α. $\|u\|_\alpha$ will denote, obviously, the α- norm of the bilinear form. The class of all linear maps of type α from E to F, endowed with the α-norm, will be denoted by $\mathcal{L}^\alpha(E, F)$, that is

$$\mathcal{L}^\alpha(E, F) := (E\widehat{\otimes}_{\alpha'} F')' \cap \mathcal{L}(E, F).$$

(Even more, $\mathcal{L}^\alpha$ is a (maximal) *normed operator ideal*, in the sense of the theory later developed by Pietsch. There is a one-to-one correspondence between maximal normed operator ideals and tensor norms, and this duality has proved to be extremely useful. (See Ref. [5] for details.)

Since $\pi' = \epsilon$ and $\epsilon' = \pi$, the linear maps of type π are precisely the *integral* maps, and those of type ϵ are *all* the continuous linear maps. Instead of defining different tensor norms and look at the corresponding classes of linear maps (what was done much later by different authors, who identified the tensor norms that produces the absolutely p-summing, p-integral, p-dominated or (p, q)-factorable operators, among many others), Grothendieck develops a general theory of tensor norms, defining new operations (the *right* and *left projective* and *injective* hull of a tensor norm, connected to factorization properties of the associated linear maps through classical Banach spaces of type $\mathcal{C} = C(K)$, $L = L_1(\mu)$ and $H =$ Hilbert space.

Grothendieck proves the fundamental result that if one starts with the ϵ-norm and takes duals, transposed, right or left injective or projective hulls finitely many times, then one obtains, up to equivalence, only 14 different tensor norms (the *natural* tensor norms), and each of them gives rise to a class of operators characterized by a typical factorization. (Ref. [11, p. 37], [5, Chapter 27])

Chapter 3 is devoted to the study of tensor norms on Hilbert spaces. The so called *hilbertian tensor norm* is introduced by the property that the corresponding linear maps factorize through a Hilbert spaces (in Ref. [5] is designed as w_2; in Ref. [7] is noted as γ_2). The relationships with other tensor norms and the different classes of operators that appear, are studied. In particular, canonical factorization results for mappings $\mathcal{C} \to H$ and $H \to L$ are obtained.

The deepest and most and influential results appear in Chapter 4: Theorem 4.1, which Grothendieck calls *théorème fondamental de la théorie métrique des produits tensoriels.* It states that the identity operator on a Hilbert space is what Grothendieck calls "preintegral", and its preintegral norm is bounded by a universal constant K_G (*Grothendieck's constant*). Grothendieck gave several other formulations (in particular, that the tensor norms π and w_2 are equivalent on $\mathcal{C} \otimes \mathcal{C}$) and obtained relevant applications to factorization of operators, Harmonic Analysis, summable sequences, etc. One of the most important achievements in the previously mentioned Lindenstrauss and Pelczynski's celebrated paper of 1968 was to realize the importance of this theorem and its reformulation in the form of an inequality involving $n \times n$ matrices and Hilbert spaces: *Let* (a_{ij}) *be an* $n \times n$ *scalar matrix. Then for each Hilbert space* H,

$$\sup \left\{ \left| \sum_{i,j=1}^{n} a_{ij}(x_i|y_j) \right| : \|x_i\|, \|y_j\| \leq 1 \right\} \leq K_G \sup \left\{ \left| \sum_{i,j=1}^{n} a_{ij} s_i t_j \right| : |s_i|, |t_j| \leq 1 \right\}$$

This is why the theorem is now called *Grothendieck's inequality.* Their proof owes much to that of Grothendieck, but this formulation, allied with the theory of p-summing operators, avoids much of the machinery of tensor products to present many of Grothendieck's ideas. Just in Ref. [15] several important results on the structure of $C(K)$ and $\mathcal{L}_p$ spaces (some of them due to Grothendieck himself!) are obtained. (See Ref. [7, Chapter 3] for a modern exposition).

As in the case of his Thesis, Grothendieck published some of the results of the *Résumé* before its appearance (Refs. [12] and [13]).

Since 2002 a series of articles under the generic title of *the metric theory of tensor products (Grothendieck's résumé revisited)* are appearing in the South African Journal *Quaestiones Math.* by **J. Diestel, J. Fourie** and **J. Swat**; for the moment five papers have been published, the most recent being in issue No. 4 of Vol. **26**, corresponding to 2003.

3. Conclusion

As we have seen, Grothendieck's life is astonishing, both personally and mathematically. His influence upon twentieth century mathematics is enormous and not only because of his magnificent results, but also for his attitude and special vision. His look for general theories and methods and the relationships between different areas of mathematics opened new fields and lines of research, sometimes developed many years after his contribution.

Appendix A. Grothendieck's Publications on Functional Analysis.

(1) *Sur le complétion du dual d'un espace vectoriel localement convexe.* C. R. Acad. Sci. Paris **230** (1950), 605-606.
(2) *Quelques résultats relatifs à la dualité dans les espaces* ($\mathcal{F}$). C. R. Acad. Sci. Paris **230** (1950), 1561-1563.
(3) *Critères généraux de compacité dans les espaces vectoriels localement convexes. Pathologie des espaces* ($\mathcal{LF}$). C. R. Acad. Sci. Paris **231** (1950), 940-941.
(4) *Quelques résultats sur les espaces vectoriels topologiques.* C. R. Acad. Sci. Paris **233** (1951), 839-841.
(5) *Sur une notion de produit tensoriel topologique d'espaces vectoriels topologiques, et une classe remarquable d'espaces vectoriels liée à cette notion.* C. R. cad. ci. Paris **233** (1951), 1556-1558.

(6) *Critères de compacité dans les espacies fonctionnels généraux.* Amer. J. of Math. **74** (1952), 168-186.

(7) *Sur les applications linéaires faiblement compactes d'espaces du type C(K).* Canadian J. Math. **5** (1953), 129-173.

(8) *Sur les espaces de solutions d'une classe générale d'équations aux dérivées partielles.* J. Analyse Math. **2** (1953), 243-280.

(9) *Sur certains espaces de fonctions holomorphes, I.* J. Reine Angew. Math. **192** (1953), 35-64.

(10) *Sur certains espaces de fonctions holomorphes, II.* J. Reine Angew. Math. **192** (1953), 77-95.

(11) *Quelques points de la théorie des produits tensoriels topologiques.* Segundo symposium sobre algunos problemas matemáticos que se están estudiando en Latino América, Julio 1954, 173-177. Centro de Cooperación Científica de la UNESCO para América Latina, Montevideo, Uruguay, 1954.

(12) *Espaces vectoriels topologiques.* Instituto de Matematica Pura e Aplicada, Universidade de São Paulo, 1954.

(13) *Résumé des résultats essentiels dans la théorie des produits tensoriels topologiques et des espaces nucléaires.* Ann. Inst. Fourier **4** (1952), 73-112.

(14) *Sur certains sous-espaces vectoriels de L^p.* Canadian J. Math. **6** (1954), 158-160.

(15) *Résultats nouveaux dans la théorie des opérations linéaires, I.* C. R. Acad. Sci. Paris **239** (1954), 577-579.

(16) *Résultats nouveaux dans la théorie des opérations linéaires, II.* C. R. Acad. Sci.Paris **239** (1954), 607-609.

(17) *Sur les espaces $(\mathcal{F})$ et $(\mathcal{D}F)$.* Summa Brazil. Math. **3** (1954), 57-123.

(18) *Produits tensoriels topologiques et espaces nucléaires.* Mem. Amer. Math. Soc. No. 16, 1955.

(19) *Une caractérisation vectorielle-métrique des espaces L^1.* Canadian J. Math. **7** (1955), 552-561.

(20) Erratum au mémoire *Produits tensoriels topologiques et espaces nucléaires.* Ann. Inst. Fourier **6** (1955-56), 117-120.

(21) *Résumé de la théorie métrique des produits tensoriels topologiques.* Bol. Soc. Mat. São Paulo **8** (1956), 1-79.

(22) *La théorie de Fredholm.* Bull. Soc. Math. France **84** (1956), 319-384.

(23) *Sur certaines classes de suites dans les espaces de Banach, et le*

théorème de Dvoretzky-Rogers. Bol. Soc. Mat. São Paulo **8** (1956), 81-110.

(24) *Un résultat sur le dual d'une C*-algèbre.* J. Mat. Pures Appl. **36** (1957), 97-108.

(25) *The trace of certain operators.* Studia Math. **20** (1961), 141-143.

References

1. L. Alonso and A. Jeremías, *La obra de Alexander Grothendieck.* La Gaceta de la RSME, vol. **4**, No **3** (2001), 623-638.
2. F. Bombal, *Sobre algunas propiedades de espacios de Banach.* Revista Acad. Ci. Madrid, **84** (1990), 83-116.
3. P. Cartier, L. Illusie, N.M. Katz, G. Laumon, Y. Manin and K. A. Ribets (ed.), *The Grothendieck Festschrift,* a collection of articles written in honor of 60th birthday of Alexander Grothendieck (3 Volumes). Progress in Mathematics, **88**. Birkhauser, Boston 1990.
4. P. Cartier, *A mad day's work: from Grothendieck to Connes and Kontsevich. The evolutions of concepts of space and symmetry.* Bull. Amer. Math. Soc. **38**, 4 (2001), 389-408.
5. A. Defant and K. Floret, *Tensor norms and Operator Ideals.* North-Holland Mathematical Studies **176**. North Holland, , Amsterdam, 1993.
6. J. Diestel, *A survey of results related to the Dunford-Pettis property.* Proc. of the Conference on Integration, Topology and Geometry in Linear Spaces,. Contemp. Math. **2** (1980), 15-60.
7. J. Diestel, H. Jarchow and A. Tongue, *Absolutely Summing Operators.* Cambridge Univ. Press, 1995.
8. G1 A. Grothendieck, *Produits tensoriels topologiques et espaces nucléaires.* Mem. Amer. Math. Soc., No. **16**, 1955.
9. A. Grothendieck, *Rèsumé des resultats essentiels dans la théorie des produits tensoriels topologiques et des espaces nucléaires.* Ann. Inst. Fourier Grenoble **4** (1952), 73-112 (appeared in 1954).
10. A. Grothendiek, *Sur les applications linéaires faiblement compactes d'espaces du type C(K).* Canadian J. Math., **5** (1953), 129-173.
11. A. Grothendieck, *Résumé de la théorie métrique des produits tensoriels topologiques.* Bol. Soc. Mat. Sao Paulo **8** (1953), 1-79 (appeared in 1956).
12. A. Grothendieck, *Résultats nouveaux dans la théorie des opérations linéaires I.* C. R. Acad. Sci. Paris **239** (1954), 577-579.
13. A. Grothendieck, *Résultats nouveaux dans la théorie des opérations linéaires II.* C. R. Acad. Sci. Paris **239** (1954), 607-609.
14. A. Grothendieck, Grothendieck on Prizes. The Mathematical Intelligencer, Vol. **11**, No. 1 (1989), 34-35.
15. J. Lindenstrauss and A. Pelczyński, *Absolutely summing operators in $\mathcal{L}_p$ spaces and applications.* Studia Mathematica, **29** (1968), 275-326.
16. L. Schwartz, *Un mathématicien aux prises avec le siècle.* Editions Odile Jacob, 1997.

THE HARDY-LITTLEWOOD MAXIMAL FUNCTION AND SOME OF ITS VARIANTS

JAVIER DUOANDIKOETXEA*

Departamento de Matemáticas,
Universidad del País Vasco-Euskal Herriko Unibertsitatea,
Apartado 644, 48080 Bilbao (Spain)
E-mail: javier.duoandikoetxea@ehu.es

The Hardy-Littlewood maximal function introduced in 1930 has several applications, the most popular being Lebesgue's differentiation theorem. The maximal function and the differentiation theorems can be generalized in higher dimensions in different ways and a review of some of them will take the first part of this paper. For the second part we present the Kakeya maximal function and maximal functions defined through sets of directions. In both cases interesting problems remain unsolved to this day.

1. Introduction

Hardy and Littlewood introduced the one-dimensional maximal function in 1930 and its higher dimensional version was first used by Wiener in 1939. Since then the operator has been widely studied and used. One of its applications is Lebesgue's differentiation theorem, which can be deduced from the boundedness of the maximal operator. The generalization of the differentiation theorem to averages over a variety of families of sets leads to the definition of several variants of the Hardy-Littlewood maximal operator. Important cases are averages over balls or cubes (the usual Hardy-Littlewood maximal function), averages over rectangles with sides parallel to the coordinate axes, and averages over arbitrary rectangles. A review of these operators and of the concept of differentiation basis will be the subject of the first part of this paper.

For the second part we haven chosen two among the many ways followed in the study of maximal operators, namely, the Kakeya maximal

*Work partially supported by grant BFM2002-01550 of MCYT (Spain) and FEDER, and by European Project HPRN-CT-2001-00273-HARP.

function and maximal functions defined through sets of directions. The size estimates of the L^p-norm of the Kakeya maximal function are useful in the study of the spherical summation multipliers for the Fourier transform (Bochner-Riesz operators) and sharp results are only known in two dimensions. For the directional operators it is possible to consider many different problems depending on the set of directions involved and we show some known results together with open questions.

The paper is an expanded version of the talk given at the Second International School of Mathematical Analysis in Andalucía. I dedicated that talk —and also this paper— to the memory of Miguel de Guzmán: scheduled as one of the main speakers of the School, he died suddenly a few months earlier. The subject of my talk had been chosen before his death and it happened to be directly related to his work and the work of some of his students in the 1970's. His two excellent monographs [22] and [23], issued from his courses and the seminar he conducted at the Universidad Complutense in Madrid, are excellent references in the field. On the other hand, his influence has been recognized as essential in the spectacular rising of Mathematics in Spain in the last thirty years; personally, I am very grateful to him for the benefit I got from his many activities.

2. The one-dimensional maximal function

Hardy and Littlewood introduced the one-dimensional maximal function in 1930 ([24]). Their definition was given for functions in an interval $(0, a)$ of the real line and was the following:

$$f^*(x) = \sup_{0<y<x} \frac{1}{x-y} \int_y^x |f(t)|dt. \tag{1}$$

It is defined for locally integrable functions and Hardy and Littlewood proved that the operator $f \mapsto f^*$ is bounded on L^p for $p > 1$; they also showed that for $p = 1$ the boundedness can fail, and that f^* is in L^1 if $f \log_+ f$ is integrable (here $\log_+ t = \max(0, \log t)$). As the title of the paper suggests they introduced the operator with some immediate applications in mind, and they said in the introduction "we have other such applications in view".

It was later observed that for $p = 1$ a more precise estimate holds, the so called *weak type* $(1, 1)$. This estimate leads to one of the usual applications of the maximal operator: Lebesgue's differentiation theorem. This theorem, published by Lebesgue in 1904, says that if f is a locally integrable function in $\mathbf{R}$, then

$$\lim_{h\to 0}\frac{1}{h}\int_x^{x+h} f(t)\,dt = f(x) \quad \text{a.e.}$$

The Hardy-Littlewood maximal function given in 1 is usually called *one-sided maximal function*, which has itself two versions, the *left-sided* and the *right-sided*. Moreover, it is defined often on the whole real line and not only on an interval. Usually other variants are considered, namely,

$$\sup_{h>0}\frac{1}{2h}\int_{x-h}^{x+h}|f(t)|\,dt,$$

$$\sup_{h,k>0}\frac{1}{h+k}\int_{x-k}^{x+h}|f(t)|\,dt,$$

which are named *centered* and *uncentered maximal function*, respectively, when there is some interest to distinguish them from each other. From the point of view of the L^p estimates all of them have a similar behavior and, unless otherwise stated, the notation Mf will correspond to any one of them.

2.1. *Strong and weak estimates*

Our first interest is size estimates of the form

$$\|Mf\|_p \le C_p\|f\|_p, \tag{2}$$

where $\|\cdot\|_p$ is the norm in $L^p(\mathbf{R})$. We observe the following.

(i) For $p=\infty$ the result is trivial: $\|Mf\|_\infty \le \|f\|_\infty$;
(ii) but for $p=1$ the result is false: indeed, $Mf(x)\sim|x|^{-1}$ when x is large if f is not identically zero (integrability can fail even locally as was shown in [24]).

A good substitute when $p=1$ is

$$|\{x : Mf(x)>\lambda\}| \le \frac{C}{\lambda}\|f\|_1 \quad \text{for}\lambda>0. \tag{3}$$

(Here and in the sequel $|E|$ will denote the Lebesgue measure of the set E.) It is good enough because 2 can be obtained for $p>1$ from this estimate together with the trivial L^∞ result (see the interpolation theorem below), and 3 is also good because Lebesgue's differentiation theorem comes as a corollary.

A norm estimate like (2) for an operator is called a *strong* (p,p) *estimate* and we say that the operator is of *strong type* (p,p), while the size estimate

$$|\{x : Mf(x) > \lambda\}| \leq \frac{C}{\lambda^p}\|f\|_p^p \quad \text{for } \lambda > 0 \tag{4}$$

is named *weak* (p,p) *estimate* and the operator is of *weak type* (p,p). It is easy to check that strong (p,p) implies weak (p,p) but the opposite is false.

The weak inequality can be read in terms of *Lorentz spaces*: (4) means that M is bounded from L^p to the Lorentz space $L^{p,\infty}$. For the definition and properties of Lorentz spaces we refer to the book [2].

Theorem 2.1. *The Hardy-Littlewood maximal operator is of weak type* $(1,1)$ *and of strong type* (p,p) *for* $1 < p < \infty$.

There are several proofs of the this theorem. The original paper by Hardy and Littlewood proves the strong type by using a limiting argument from the analogous theorem for finite sums and decreasing rearrangements. Shortly after (1931) R. M. Gabriel published a simpler proof that inspired a beautiful proof of F. Riesz (1932) in which he used his "rising sun lemma" to obtain the weak inequality. We refer to [44] or [2] for the use of rearrangements; the proof of F. Riesz appears in [22], [16] and [44], for instance. The rising sun lemma gives for any of the one-sided versions of the maximal operator (3) the following *equality*:

$$|\{x : Mf(x) > \lambda\}| = \frac{1}{\lambda}\int_{\{x:Mf(x)>\lambda\}} |f|. \tag{5}$$

A different type of proof by using a covering lemma will be better adapted to higher dimensions. The convenient form in the one-dimensional case is the following.

Lemma 2.1. *Given a finite family of intervals of the real line, there is a subfamily with the same union and such that each point is at most in two intervals of the subfamily.*

The lemma is easily deduced from the following observation: *if all the intervals of a finite family have a common point, two of them are enough to cover the union of the whole family.*

With this lemma the proof of the weak $(1,1)$ estimate is as follows. If $Mf(x) > \lambda$, then there exists an interval $I(x)$ containing x such that

$$\frac{1}{\lambda}\int_{I(x)} |f| > |I(x)|.$$

Let K be a compact set contained in $E_\lambda = \{x : Mf(x) > \lambda\}$; there is a finite covering of K with such intervals. By the lemma there is a subcovering $\{I_j\}$ such that $\sum \chi_{I_j}(x) \le 2$ (χ_I is the characteristic function of I). Then

$$|K| \le \sum_j |I_j| \le \frac{1}{\lambda} \sum_j \int_{I_j} |f| \le \frac{1}{\lambda} \int \sum_j \chi_{I_j} |f| \le \frac{2}{\lambda} \int |f|.$$

The formula

$$\int |g|^p = \int_0^\infty p\lambda^{p-1} |\{x : |g(x)| > \lambda\}| \, d\lambda$$

is useful to insert the weak $(1,1)$ estimate into the L^p norm of Mf. This can be stated as an interpolation theorem.

Theorem 2.2. *Let T be a sublinear operator satisfying a weak type $(1,1)$ estimate and bounded on L^∞. Then it is bounded on L^p for $1 < p < \infty$.*

(Sublinear means that $|T(f+g)| \le |Tf| + |Tg|$.) A proof of the theorem can be found in [16] or with many other variants in [2]. More generally, it can be proved that if T is of weak type (p_0, p_0) and of weak type (p_1, p_1), then it is of strong type (p,p) for $p_0 < p < p_1$.

2.2. *The differentiation theorem*

To conclude the one-dimensional case we show how to deduce Lebesgue's differentiation theorem from the weak type $(1,1)$ for M.

First we notice the following inequalities:

$$\left| \limsup_{h \to 0} \frac{1}{h} \int_x^{x+h} f(t)\, dt - \liminf_{h \to 0} \frac{1}{h} \int_x^{x+h} f(t)\, dt \right| \le 2Mf(x), \tag{6}$$

and

$$\left| \lim_{h \to 0} \frac{1}{h} \int_x^{x+h} f(t)\, dt - f(x) \right| \le Mf(x) + |f(x)|. \tag{7}$$

Let $\Lambda f(x)$ be any one of the left-hand sides of the inequalities; for continuous g, $\Lambda g(x)$ is zero and $\Lambda f(x) = \Lambda(f-g)(x)$. Let $f \in L^1$ and $\epsilon > 0$, and choose a continuous g such that $\|f - g\|_1 < \epsilon$; then

$$|\{x : \Lambda f(x) > \frac{1}{n}\}| = |\{x : \Lambda(f-g)(x) > \frac{1}{n}\}| \le Cn\epsilon,$$

by using the weak $(1,1)$ estimate for M. Since this holds for all $\epsilon > 0$ we conclude that the measure of $\{x : \Lambda f(x) > 1/n\}$ has to be zero, as will be the measure of the union of those sets for all $n \in \mathbf{N}$.

When $\Lambda f(x)$ is the left-hand side of (6) we deduce that the limit of $h^{-1}\int_x^{x+h} f$ exists for almost every x; with the left-hand side of (7) we conclude that it equals $f(x)$ almost everywhere. Although the proof needs f integrable, the result is local and the conclusion holds for locally integrable f.

For a discussion on the equivalence of convergence, finiteness and size estimates see the books of Guzmán [23] or Stein [36].

3. The maximal function in higher dimensions

In 1910 Lebesgue extended to higher dimensions his differentiation theorem by considering averages on Euclidean balls. The corresponding extension to $\mathbf{R}^n$ of the Hardy-Littlewood maximal function was introduced by N. Wiener in [40] in a different context. The definition is

$$Mf(x) = \sup_{r>0} \frac{1}{|B(x,r)|} \int_{B(x,r)} |f(y)|\, dy.$$

As in the one dimensional case this definition corresponds to the *centered maximal function* and for the *uncentered* we require the balls simply to contain the point. Although this defines a bigger operator, it is equivalent in size to the centered one because the uncentered ball is contained in a centered ball with double radius.

More generally, one can start with a fixed set B containing the origin and define a maximal function by considering all the family of sets obtained using dilations and translations of B:

$$M_B f(x) = \sup_{r>0} \frac{1}{|rB|} \int_{rB} |f(x+y)|\, dy.$$

If there are two balls centered at the origin with radii r_1 and r_2 such that $B(0,r_1) \subset B \subset B(0,r_2)$, then M_B is equivalent to M in the sense that the quotient $Mf(x)/M_B f(x)$ is bounded above and below by constants depending only on r_1, r_2, and the dimension, and not on f or x. In particular, this is true when B is the ball defined by an l^q-norm in $\mathbf{R}^n$.

Theorem 2.1 holds true for the higher dimensional M and, of course, its equivalent variants. The L^∞ estimate is trivial again and the key estimate is the weak type $(1,1)$. It is clear that Lemma 2.1 is false in higher dimensions: indeed, we cannot always select from three overlapping balls two of them in such a way that their union covers the third one. Then one needs to find some substitute for that lemma. Wiener's way [40] is a version of Vitali lemma.

Lemma 3.1. *Let $\{B_j\}_{j\in\mathcal{J}}$ be a collection of balls in $\mathbf{R}^n$. Then there exists an at most countable subcollection of disjoint balls $\{B_k\}$ such that*

$$\bigcup_{j\in\mathcal{J}} B_j \subset \bigcup_k 5B_k.$$

Here the selected balls are disjoint but they do not cover the original union. A proof of the lemma can be seen in [36]. Closer in spirit to Lemma 2.1 but harder to prove is the next result due independently to Besicovitch and Morse.

Lemma 3.2. *Let A be a bounded set in $\mathbf{R}^n$, and suppose that $\{B_x\}_{x\in A}$ is a collection of balls such that $B_x = B(x, r_x)$, $r_x > 0$. Then there exists an at most countable subcollection of balls $\{B_j\}$ and a constant C_n, depending only on the dimension, such that*

$$A \subset \bigcup_j B_j \quad \textit{and} \quad \sum_j \chi_{B_j}(x) \le C_n.$$

With any one of these lemmas we can obtain the weak type $(1,1)$ of the maximal operator if we follow the proof given in the one-dimensional case. In a similar way the differentiation theorem comes as a consequence.

More results on covering lemmas are in the book by M. de Guzmán [22].

3.1. *Directional maximal functions*

Starting from the one-dimensional operator we can define directional operators in $\mathbf{R}^n$ as follows. Let $u \in S^{n-1}$ be a direction in $\mathbf{R}^n$; for smooth f define the directional Hardy-Littlewood maximal operator associated to u as

$$M_u f(x) = \sup_{h>0} \frac{1}{h} \int_0^h |f(x+tu)|\, dt. \tag{8}$$

In particular, the operator associated to the vector e_j of the standard basis of $\mathbf{R}^n$ will be represented by M_j (instead of M_{e_j}) and corresponds to

$$M_j f(x) = \sup_{h>0} \frac{1}{h} \int_0^h |f(x_1, \ldots, x_j + t, \ldots, x_n)|\, dt.$$

By using the one-dimensional Theorem 2.1 and Fubini's theorem we deduce that *M_u is bounded on L^p for $1 < p \le \infty$ and is of weak type $(1,1)$ with the constants of the one-dimensional operator*, that is, with constants independent of u.

Let B_R be the Euclidean ball centered at x with radius R. If we use polar coordinates in $\mathbf{R}^n$ we can write

$$\begin{aligned}\frac{1}{R^n|B_1|}\int_{B_R}|f(x-y)|\,dy &= \frac{1}{|B_1|}\int_{S^{n-1}}\frac{1}{R^n}\int_0^R |f(x-tu)|t^{n-1}\,dt\,d\sigma(u)\\ &\leq \frac{1}{|B_1|}\int_{S^{n-1}} M_u f(x)\,d\sigma(u).\end{aligned}$$

By taking the supremum for $R > 0$ we see that the last term is a pointwise bound for the centered maximal function. By using now the uniform L^p boundedness of M_u and Minkowski integral inequality we obtain a new proof of the L^p boundedness of M for $1 < p < \infty$. (It does not work for the weak type $(1,1)$.) Moreover, since the quotient of the measures of the unit sphere and the unit ball is n we get a bound for the norm of the operator of $A_p n$ (this, in turn, can be lowered to a multiple of $n^{1/p}$ by interpolating with the L^∞ estimate). This was observed by M. de Guzmán (see the books [22] and [23]). It is a remarkable fact that such a simple proof gives a much better size for the constant of the centered Hardy-Littlewood maximal operator than the usual proofs through covering lemmas from which one obtains an exponential growth in n. It should be clear that we obtain this size only for the centered maximal operator on Euclidean balls and that the equivalence of the several definitions from the point of view of L^p-boundedness does not extend to the size of the norms.

Actually, it can be seen that the size of the L^p operator norm for the centered maximal operator on balls is independent of the dimension for $1 < p \leq \infty$ (the result was announced by E. Stein and appeared published in [38]). In the one-dimensional case some attention has been paid to the exact value of the constants: L. Grafakos and S. Montgomery-Smith obtained the exact L^p-operator norm of the uncentered maximal operator in [21] and A. Melas gave the sharp weak $(1,1)$ constant for the centered case in [32] (for the uncentered it is immediately seen to be 2).

4. The strong maximal function

Instead of balls or cubes we can use in $\mathbf{R}^n$ many other families of sets. A natural extension of one-dimensional intervals are the Cartesian products of such intervals. Those product sets will be called *rectangles* as well as the sets obtained by rotating them. The direction of the largest side of a rectangle will be called the direction of the rectangle. (If several sides are of the same length any one of them will serve to define the direction.) Let $\mathcal{R}_n$ be the set of rectangles in $\mathbf{R}^n$ with sides parallel to the coordinate axes.

The *strong maximal function* is defined as

$$M_S f(x) = \sup_{x \in R \in \mathcal{R}_n} \frac{1}{|R|} \int_R |f(y)|\, dy.$$

This means that we take the supremum of averages over all rectangles containing x and with sides parallel to the axes.

If we take as f the characteristic function of the unit ball, for instance, then we can check that when the coordinates $x_1, \dots, x_n$ of x are large, the size of $M_S f(x)$ is of the order $(|x_1| \cdot \ldots \cdot |x_n|)^{-1}$ so that the weak type $(1,1)$ inequality cannot be satisfied. On the other hand, $M_S f(x)$ is always pointwise bounded by the composition $M_1(M_2(\dots(M_n f)\dots))$ where the M_j are the directional operators defined in the previous section. Each one of these operators is bounded on $L^p(\mathbf{R}^n)$ so that M_S is also bounded on $L^p(\mathbf{R}^n)$.

As a corollary we can say that the differentiation holds for rectangles with sides parallel to the coordinate axes if $f \in L^p_{\text{loc}}$ for some $p > 1$. This differentiation result had been proved in the 1930's by A. Zygmund, without using the maximal operator. It was improved by Jessen, Marcinkiewicz and Zygmund who showed that it is enough to have f locally in $L(\log L)^{n-1}$; Saks proved that the differentiation result can fail if $p = 1$.

To obtain the theorem of Jessen, Marcinkiewicz and Zygmund through the strong maximal function we will need to find an appropriate substitute for the weak $(1,1)$ inequality; the following theorem shows it.

Theorem 4.1. *The strong maximal function satisfies the following inequality for $\lambda > 0$,*

$$|\{x : M_S f(x) > \lambda\}| \le C \int_{\mathbf{R}^n} \frac{|f(x)|}{\lambda} \log\left(1 + \frac{|f(x)|}{\lambda}\right)^{n-1} dx. \tag{9}$$

Let A be a measurable set of $\mathbf{R}^n$. The space $L \log^k L(A)$ is the set of measurable functions f such that

$$\int_A |f(x)|(1 + \log_+ |f(x)|)^k\, dx < \infty.$$

(Here $\log_+ t = \max(0, \log t)$.) If A is of finite measure, then this space contains $L^p(A)$ for all $p > 1$. The space $L \log^k L(A)$ is one of the so-called *Orlicz spaces.* See the book of Zygmund [44] or the monography by M. M. Rao and Z. D. Ren [34].

The quantitative result (9) is due to M. de Guzmán. A geometrical proof that uses appropriate covering lemmas is due to A. Córdoba and R. Fefferman [14]. It is reproduced as an appendix in the book [22].

5. The universal maximal function

The definition of the strong maximal functions depends on the choice of the axes in the coordinate system. By using all rectangles in all directions we get a much bigger maximal function. It is sometimes called *universal maximal function.* The definition is

$$\mathcal{M}f(x) = \sup_{x \in R} \frac{1}{|R|} \int_R |f(y)| \, dy,$$

where R can be any rectangle containing x, regardless of direction.

If we choose again as f the characteristic function of the unit ball, then the size of $\mathcal{M}f(x)$ for big x becomes like $|x|^{-1}$; such a function is in L^p only for $p > n$, and as a consequence $\mathcal{M}$ is unbounded for $p \leq n$. Actually, a much deeper result holds.

Theorem 5.1. *$\mathcal{M}$ is unbounded on L^p for all finite p.*

The counterexample is more involved. It is based on a construction due to Besicovitch which leads to the following situation: given $\epsilon > 0$, there exists a set E such that $|E| < \epsilon$ and there is a set $\tilde{E}$ of measure greater than 1 such that $\mathcal{M}\chi_E(x) \geq 1/2$ for $x \in \tilde{E}$. Details of the construction of such a set E by using the so-called *Perron tree* can be seen in the books [15], [22], [23], and [36], for instance.

From the differentiation point of view this would correspond to a negative result when approaching a point by using rectangles with arbitrary direction. This negative result can hold even for characteristic functions, a result due to Zygmund.

Since the characteristic function of the ball does not provide us with a counterexample for $p > n$, this suggests that a positive result could be possible if we restrict to radial functions. This is the content of the following theorem.

Theorem 5.2. *$\mathcal{M}$ restricted to radial functions is bounded on L^p for $p > n$ and is of restricted weak type (n, n).*

The last part of the statement means that it satisfies a weak (n, n) estimate for characteristic functions of sets, that is,

$$\lambda^n |\{x : \mathcal{M}\chi_E(x) > \lambda|\} \leq C|E|, \tag{10}$$

when E is radial. As the usual weak (p, p) inequality the restricted one can be read in terms of Lorentz spaces: (10) is equivalent to the boundedness of $\mathcal{M}$ from $L^{n,1}$ (restricted to radial functions) to $L^{n,\infty}$.

This theorem was proved by A. Carbery, E. Hernández and F. Soria and can be seen in [7]. An alternative proof was given in [18], where the following pointwise inequality was obtained: for radial sets E,

$$\mathcal{M}\chi_E(x) \leq C(M\chi_E(x))^{1/n} \quad \text{a.e.}$$

From it and the weak type $(1,1)$ of M, (10) is immediate. The result for $p > n$ can be obtained by interpolation with L^∞ (see [2]). In [17] similar results are obtained for l^q-radial functions.

6. Differentiation bases

In a more general way one can consider differentiation bases and maximal functions assocated to them. A *differentiation basis* in $\mathbf{R}^n$ is a family $\mathcal{B}$ of sets of $\mathbf{R}^n$ with the following property: for each x there is a subfamily $\mathcal{B}(x)$ of sets of $\mathcal{B}$ associated to x; each set in $\mathcal{B}(x)$ contains x, and there is a sequence of sets of $\mathcal{B}(x)$ with diameters tending to zero. A differentiation basis is a *Buseman-Feller differentiation basis* if all its sets are open and if $\mathcal{B}(x)$ is the family of all sets in $\mathcal{B}$ containing x. To a differentiation basis we can associate a maximal operator

$$M_{\mathcal{B}}f(x) = \sup_{x \in B \in \mathcal{B}} \frac{1}{|B|} \int_B |f(y)|\, dy.$$

The classical Hardy-Littlewood maximal operator, the strong maximal function, and the universal maximal function correspond to maximal operators associated to several differentiation bases. The centered versions do not correspond to Buseman-Feller bases.

We say that the basis $\mathcal{B}$ *differentiates* f if

$$\lim \left\{ \frac{1}{|B|} \int_B f(y)\, dy : \delta(B) \to 0, B \in \mathcal{B}(x) \right\} = f(x) \quad \text{a.e.}$$

(Here $\delta(B)$ denotes the diameter of B.) The basis is said to be a *density basis* if it differentiates characteristic functions of measurable sets.

There are intimate relations among differentiation properties, boundedness of the maximal operator, and covering theorems for a basis. We will show a sample theorem (see the papers [13] and [27]) and refer to the aforementioned books of M. de Guzmán for more results.

To state the theorem we need two definitions. Given a measurable set E in $\mathbf{R}^n$, we say that $V \subset \mathcal{B}$ is a $\mathcal{B}$-Vitali covering of E if, for every $x \in E$, there is a sequence included in V that converges to x.

A differentiation basis $\mathcal{B}$ has the covering property V_q if there exists a constant C such that, for every bounded measurable set E, every $\mathcal{B}$-Vitali covering V of E and any $\epsilon > 0$, one can select a countable set $\{R_k\} \subset V$ such that $|E - \bigcup R_k| = 0$, $|\bigcup R_k - E| \leq \epsilon$, and $\|\sum \chi_{R_k}\|_q \leq C|E|^{1/q}$.

Theorem 6.1. *Let $\mathcal{B}$ be a differentiation basis invariant under translations; then $\mathcal{B}$ differentiates all $f \in L^p_{loc}(R^n)$ for some $1 < p \leq \infty$ if and only if it has the covering property $V_{p'}$ with $1/p + 1/p' = 1$.*

7. Some further developments

One of the main achievements in the study of maximal functions was the characterization of weighted inequalities for the Hardy-Littlewood maximal operator (and for the strong maximal function), that is, the description of the measures μ such that M is bounded from $L^p(\mu)$ to $L^p(\mu)$ or is of weak type (p,p) for the spaces defined with the measure μ. The characterization is due to B. Muckenhoupt and is as follows: μ has to be absolutely continuous with respect to the Lebesgue measure, and its density w (that is, $d\mu(x) = w(x)dx$) must satisfy a condition given in terms of averages of w and $w^{1-p'}$ (*A_p condition*). The fact that the classes of weights satisfying such conditions have a rich structure led to a wide development in the study of weighted inequalities. Moreover, it happened that singular integral operators with smooth kernel are also bounded in the same weighted spaces. The basic theory is very well established and can be found in several books; since we will not enter into its details we refer the interested reader to the books [16] and [36], for instance.

Another development came from the consideration of averages over lower dimensional sets; for instance, spheres centered at the point or portions of curves. We limit ourselves to mention the *spherical maximal operator*. It consists on taking as the maximal function in each point the supremum of the averages of the function over all the spheres centered at it. The corresponding operator is bounded on L^p if and only if $p > n/(n-1)$. This was first proved by E. Stein for $n \geq 3$ [37] and by J. Bourgain [3] for $n = 2$. The lack of L^2-boundedness makes more difficult the two-dimensional case. Different proofs were given later for both results; see [36] and the references therein. Curiously, if the supremum is taken only on dyadic radii, that is, 2^j for $j \in \mathbf{Z}$, the corresponding operator is bounded on L^p for all $p > 1$ (there are several proofs and the first ones are due to C. Calderón and to R. Coifman and G. Weiss; another proof appears in [16]).

Other types of *rough* maximal operators came into play, but we will not even mention them, and for the continuation of the paper we have selected two topics: (a) the Kakeya maximal operator, and (b) maximal operators defined by sets of directions.

8. The Kakeya maximal operator

8.1. *The size conjecture*

Given $N > 2$, define $\mathcal{K}_N$ as the family of all rectangles with $n-1$ sides of the same length and one side of length N times bigger, that is, $h \times h \times \cdots \times Nh$, for any $h > 0$. The associated maximal function will be called *Kakeya maximal operator* and is as follows:

$$K_N f(x) = \sup_{x \in B \in \mathcal{K}_N} \frac{1}{|B|} \int_B |f|.$$

Since each rectangle has measure Nh^n and it can be included in an Euclidean ball of radius its diameter, we have the following pointwise inequality

$$K_N f(x) \leq C_n N^{n-1} M f(x). \tag{11}$$

An immediate consequence is that K_N is bounded on L^p for all $p > 1$ and is of weak type $(1,1)$. Then it is clear that the interesting questions regarding K_N will not concern its boundedness. Indeed, the problem is to study the growth of the L^p operator norm of K_N as a function of N. If we use N^{n-1} as the size of the weak $(1,1)$ constant for K_N (deduced from (11)) and then interpolate with the trivial L^∞ estimate, we obtain the growth $N^{(n-1)/p}$ for the norm in L^p. But, as the reader already guessed, this will not be sharp.

Conjecture. For $1 < p < \infty$ we have $\|K_N f\|_p \leq C(p,N)\|f\|_p$ and

$$C(p,N) \leq \begin{cases} C(\log N)^{\alpha(p)}, & \text{if } p \geq n; \\ CN^{n/p-1}(\log N)^{\alpha(p)}, & \text{if } 1 < p < n. \end{cases} \tag{12}$$

The exponent $\alpha(p)$ is nonnegative and it must be strictly positive for $p \geq n$.

8.2. *Bochner-Riesz multipliers*

One of the reasons which makes this problem interesting is its relation to the study of the boundedness of Bochner-Riesz multipliers (spherical summability of the Fourier transform). By using the Fourier transform we can define these operators as

$$(S_\alpha f)\hat{}(\xi) = (1 - |\xi|^2)_+^\alpha \hat{f}(\xi),$$

where $t_+ = \max(t, 0)$, so that the Fourier multiplier of S_α is supported in the unit ball, and for $\alpha = 0$ it coincides with the characteristic function of such unit ball. A celebrated result of C. Fefferman states that S_0 is unbounded for $p \neq 2$; his counterexample is based on the Besicovitch set that provided the counterexample for the universal maximal operator in Section 5. The original proof was published in [20] and can be found also in the books [15], [23], and [36].

For $\alpha > (n-1)/2$ the operator S_α has an integrable kernel so that it is bounded on L^p for $1 \leq p \leq \infty$. In the remaining range, $0 < \alpha \leq (n-1)/2$, the values of p for which the operator S_α is expected to be bounded are

$$\frac{2n}{n+1+2\alpha} < p < \frac{2n}{n-1-2\alpha}. \tag{13}$$

For $n = 2$ this was proved first by L. Carleson and P. Sjölin [8]. Different proofs due to L. Hörmander, C. Fefferman, and A. Córdoba appeared in the 1970's. The proof of Córdoba connected the Bochner-Riesz multiplier to the Kakeya maximal function and was based on the critical estimate for $n = p = 2$ (see [10]). Then we have the following theorem.

Theorem 8.1. *The aforementioned conjecture for the norm of K_N holds true for $n = 2$.*

In dimension higher than two the conjecture has not been fully proved to this day. In the same sense, the boundedness of the Bochner-Riesz multipliers has been established for the sharp range (13) when $\alpha > a(n)$ for some $a(n)$ strictly positive if $n \geq 3$.

The fact that the critical value in the conjecture is $p = n$ makes things easier for $p = 2$. In the same way, in higher dimensions, the correct growth of the norm can be easily obtained for $1 < p \leq 2$; even up to $p = (n+1)/2$ it does not require considerable effort. After many years without any progress, a remarkable improvement was obtained by J. Bourgain in [4]; subsequently, T. Wolff showed that the conjecture for the Kakeya maximal operator holds true in all dimensions for the range $1 < p \leq (n+2)/2$ [41]. Further improvements are due to J. Bourgain [5], N. Katz, and T. Tao. See the presentations of T. Wolff in [42] and [43], and the survey of Katz and Tao in [30] for more detailed results and references. For recent improvements on the Bochner-Riesz multipliers see also the article [31].

8.3. *The dimension conjecture*

The Kakeya conjecture is also related to the following problem in Geometric Measure Theory. Let E be a set of $\mathbf{R}^n$ containing a segment of length 1 in each direction; E might be of zero Lebesgue measure but it is conjectured that its Hausdorff dimension must be n. The relation between this and the norm size conjecture is the following: *if the conjecture for the Kakeya maximal function holds true for some $p_0 \leq n$, then the Hausdorff dimension of E is at least p_0.* For other results on dimension (including Minkowski dimension, a concept that we will introduce in the last section of the paper) the reader can consult the survey [30]. In recent years this field has progressed so quickly that it is difficult to keep track of the current state of affairs and to say which is the best exponent for each one of the conjectures at this moment.

8.4. *Radial functions*

It is clear that the Kakeya operator is pointwise bounded by the universal maximal operator. Then, as a consequence of Theorem 5.2, we deduce that the norm of K_N on L^p acting on radial functions is independent of N for $p > n$. We obtain the constant $C \log N$ for $p = n$ by interpolation. Similar results for l^q-radial functions are in [17].

9. Maximal operators on sets of directions

Let Σ be a subset of the unit sphere in $\mathbf{R}^n$. Associated to Σ we define the maximal operator

$$M_\Sigma f(x) = \sup_{u \in \Sigma} M_u f(x),$$

where M_u is the directional operator defined in (8). We remark that these operators are equivalent to the operator defined through a differentiation basis formed by all rectangles with directions in Σ. In particular, when $\Sigma = S^{n-1}$ we get the universal maximal function.

Depending on the set Σ two different kinds of questions arise:

- if Σ is finite, how does the norm of M_Σ as a bounded operator on L^p grow in terms of the cardinality of Σ?
- if Σ is infinite, is M_Σ a bounded operator on L^p?

9.1. *Finite sets of directions*

The problem for the Kakeya maximal function can be reduced to consider rectangles whose directions are in some finite set Σ provided that for any $u \in S^{n-1}$ there is a direction $v \in \Sigma$ such that the distance between u and v is less than $1/N$. In such a case we can replace each one of the rectangles in the definition of K_N with direction u by another rectangle of comparable size containing it and with direction v. The number of directions needed for such a Σ would be of the order of N^{n-1}, and they will be essentially equidistributed in S^{n-1}. Then the conjecture for M_Σ when $\operatorname{card}\Sigma = N^{n-1}$ is the same as the conjecture for K_N. In other words, we expect the norm of M_Σ to have a logarithmic growth for $p \geq n$ and to behave as $(\operatorname{card}\Sigma)^{1/p-1/n+\epsilon}$ when $1 < p < n$ ($\epsilon = 0$ can be chosen if we drop some logarithms instead). Moreover, the conjecture is that this should be the behavior regardless the fact that the directions are equidistributed or not.

Due to its relation with K_N it is clear that the answer to these questions remains open for $n \geq 3$. For $n = 2$ and equidistributed directions the result was proved by A. Córdoba in [11], and by J.-O. Strömberg in [39]. Avoiding the equidistribution hypothesis had to wait for almost twenty years until finally N. Katz solved it in [29]; an alternative and easier proof can be found in [1].

In three or more dimensions it is not difficult to prove the conjecture for $1 < p \leq 2$, and we do not know about any result for $p > 2$.

9.2. *Infinite sets of directions*

When Σ is infinite we can assume that it is closed because an approximation argument gives the result for the closure of Σ whenever it is known for Σ.

We already know that the operator is unbounded for all finite p if $\Sigma = S^{n-1}$. The same is true when Σ has positive measure. Then we only need to consider sets with zero measure.

It is possible to construct sets of zero measure for which M_Σ is unbounded. Consider $\mathbf{R}^2$, for simplicity, so that the directions are in the unit circle and can be ordered starting with one of them. Assume that the directions in a subset Σ_0 of Σ are almost equidistributed (this means that the distance between two consecutive directions is bounded above and below by absolute constants); then the norm of the maximal operator defined by Σ_0 is at least a power of $\log \operatorname{card}\Sigma_0$ for $p \geq 2$. If we can find a sequence of subsets of Σ, each one with almost equidistributed directions, and with cardinalities tending to infinity, then M_Σ will be unbounded. Defining directions in S^1 through the slope of lines, put $\Sigma(\gamma) = \{k^{-\gamma}, k \in \mathbf{N}\}$, for $\gamma > 0$, and take the values of k between 2^j and 2^{j+1} to get 2^j almost equidistributed directions. As a consequence $M_{\Sigma(\gamma)}$ is unbounded for finite p. Notice that these $E(\gamma)$ are not only of zero measure but also of zero Hausdorff dimension.

Some positive results are known: Nagel, Stein and Wainger proved in [33] that for lacunary sets of directions (like $\{2^j : j \in \mathbf{N}\}$ in the plane) the associated operator is bounded on L^p for all $p > 1$, so improving previous partial results. This was extended to more general sets by P. Sjögren and P. Sjölin [35] by inserting lacunary directions between lacunary directions.

Those were the only known results for many years. Then N. Katz came with a negative answer for the ternary Cantor set; in that case, M_Σ is unbounded on L^2 [28]. Later, K. Hare extended this negative result to all Cantor type sets of positive Hausdorff dimension [25] and the same author with J.-O. Rönning gave a necessary condition which permit to conclude unboundedness in more cases [26]. Nevertheless, the answer for $p > 2$ remains unknown for the ternary Cantor set.

We can summarize by saying that we know complete results only of two types: *either M_Σ is bounded for all $p > 1$, or it is unbounded for all finite p.* To state this as a plausible conjecture would need stronger support from more complicated sets of directions.

In higher dimensions we can construct negative results by following the same pattern as in the two-dimensional case, and positive results are known for sets of directions laying on a curve and having some kind of lacunary behavior (actually, the result proved in [33] is of this type).

The problem is better understood when restricted to radial functions. But the range of boundedness can be larger and so it is in many cases. We already knew this from the universal maximal function according to the results mentioned in Section 5.

The range of boundedness happens to be related to the *Minkowski dimension* of Σ. We recall its definition. Given a bounded set E in $\mathbf{R}^n$, let $N(\delta)$ be the minimum number of balls of radius δ needed to cover E; then the Minkowski dimension of E is defined as

$$\dim_M E = \limsup_{\delta \to 0} \frac{\log N(\delta)}{-\log \delta}.$$

Then $\dim_M E$ comes as a number in the range $[0, n]$ (not larger than $n-1$ for subsets of S^{n-1}). The Minkowski dimension is an upper bound for the Hausdorff dimension and sometimes they coincide, for Cantor sets, for instance. But, on the other hand, the set $\Sigma(\gamma)$ introduced above has Minkowski dimension $(\gamma+1)^{-1}$ while its Hausdorff dimension is 0.

The following theorem was proved in [19].

Theorem 9.1. *Let $\Sigma \subset S^1$ and $d = dim_M \Sigma$. Then M_Σ restricted to radial functions is bounded on $L^p(\mathbf{R}^2)$ if $p > 1+d$, and unbounded if $p < 1+d$.*

The case $p = 1+d$ does not depend only on the dimension of Σ and in some cases the operator is bounded, in some others it is not.

The proof of the theorem can be easily adapted to higher dimensions when $0 < d < 1$, and we also know that it holds for $d = n-1$ due to Theorem 5.2. The conjecture that Theorem 9.1 should be valid for all dimensions seems to lay on a good basis.

Going back to the general situation, we see that all the sets for which boundedness for some p has been proved are of zero Minkowski dimension. Whether there is any Σ with positive Minkowski dimension for which the operator M_Σ is bounded on L^p for some p is an open and interesting question, but maybe it is out of reach for the moment. On the other hand, we can construct a set Σ such that $\dim_M \Sigma = 0$ and M_Σ is unbounded for all finite p: take the dyadic set $\{2^{-k} : k \in \mathbf{N}\}$ and insert k equidistributed directions between 2^{-k} and 2^{-k-1}.

References

1. A. Alfonseca, F. Soria and A. Vargas, *A remark on maximal operators along directions in* $\mathbf{R}^2$, Math. Res. Lett. 10 (2003), 41–49.
2. C. Bennet and R. Sharpley, *Interpolation of Operators*, Pure and Applied Mathematics, 129. Academic Press, Inc., Boston, MA, 1988.
3. J. Bourgain, *Averages in the plane over convex curves and maximal operators*, J. Analyse Math. 47 (1986), 69–85.
4. J. Bourgain, *Besicovitch type maximal operators and applications to Fourier analysis*, Geom. Funct. Anal. 1 (1991), 147–187.
5. J. Bourgain, *On the dimension of Kakeya sets and related maximal inequalities*, Geom. Funct. Anal. 9 (1999), 256–282.
6. A. M. Bruckner, *Differentiation of integrals*, Amer. Math. Monthly 78 (1971), no. 9, Part II.
7. A. Carbery, E. Hernández and F. Soria, *Estimates for the Kakeya maximal operator on radial functions in* $\mathbf{R}^n$, Harmonic analysis (Sendai, 1990), 41–50, ICM-90 Satell. Conf. Proc., Springer, Tokyo, 1991.
8. L. Carleson and P. Sjolin, *Oscillatory integrals and a multiplier problem for the disc*, Studia Math. 44 (1972), 287–299.
9. A. Córdoba, *On the Vitali covering properties of a differentiation basis*, Studia Math. 57 (1976), 91–95.

10. A. Córdoba, *The Kakeya maximal function and the spherical summation multipliers*, Amer. J. Math. 99 (1977), 1–22.
11. A. Córdoba, *The multiplier problem for the polygon*, Ann. of Math. 105 (1977), 581–588.
12. A. Córdoba, *Geometric Fourier analysis*, Ann. Inst. Fourier (Grenoble) 32 (1982), 215–226.
13. A. Córdoba and R. Fefferman, *On differentiation of integrals*, Proc. Nat. Acad. Sci. U.S.A. 74 (1977), 2211–2213.
14. A. Córdoba and R. Fefferman, *A geometric proof of the strong maximal theorem*, Ann. of Math. 102 (1975), 95–100.
15. K. M. Davis and Yang-Chun Chang, *Lectures on Bochner-Riesz means*, London Mathematical Society Lecture Note Series, 114, Cambridge University Press, Cambridge, 1987.
16. J. Duoandikoetxea, *Fourier Analysis*, Graduate Texts in Mathematics no. 29, American Mathematical Society, Providence, RI, 2001.
17. J. Duoandikoetxea and V. Naibo, *The universal maximal operator on special classes of functions*, Indiana Univ. Math. J. 54 (2005), 1351–1370.
18. J. Duoandikoetxea, V. Naibo and O. Oruetxebarria, *k-plane transforms and related operators on radial functions*, Michigan Math. J. 49 (2001), 265–276.
19. J. Duoandikoetxea and A. Vargas, *Directional operators and radial functions on the plane*, Ark. Mat. 33 (1995), 281–291.
20. C. Fefferman, *The multiplier problem for the ball*, Ann. of Math. 94 (1971), 330–336.
21. L. Grafakos and S. Montgomery-Smith, *Best constants for uncentred maximal functions*, Bull. London Math. Soc. 29 (1997), 60–64.
22. M. de Guzmán, *Differentiation of Integrals in $\mathbf{R}^n$*, Lecture Notes in Math. 481, Springer-Verlag, 1975.
23. M. de Guzmán, *Real variable methods in Fourier analysis*. North-Holland Mathematics Studies, 46. North-Holland Publishing Co., Amsterdam-New York, 1981.
24. G. H. Hardy and J. E. Littlewood, *A maximal theorem with function-theoretic applications*, Acta Math. **54** (1930), 81–116.
25. K. E. Hare, *Maximal operators and Cantor sets*, Canad. Math. Bull. 43 (2000), 330–342.
26. K. E. Hare and J.-O. Rönning, *The size of* Max(p) *sets and density bases*, J. Fourier Anal. Appl. 8 (2002), 259–268.
27. C. A. Hayes *Derivation of the integrals of $L^{(q)}$-functions*, Pacific J. Math. 64 (1976), 173–180.
28. N. H. Katz, *A counterexample for maximal operators over a Cantor set of directions*, Math. Res. Lett. 3 (1996), 527–536.
29. N. H. Katz, *Maximal operators over arbitrary sets of directions*, Duke Math. J. 97 (1999), 67–79.
30. N. H. Katz and T. Tao, *Recent progress on the Kakeya conjecture*, Proceedings of the 6th International Conference on Harmonic Analysis and Partial Differential Equations (El Escorial, 2000). Publ. Mat. 2002, Vol. Extra, 161–179.

31. S. Lee, *Improved bounds for Bochner-Riesz and maximal Bochner-Riesz operators*, Duke Math. J. 122 (2004), 205–232.
32. A. D. Melas, *The best constant for the centered Hardy-Littlewood maximal inequality*, Ann. of Math. (2) 157 (2003), 647–688.
33. A. Nagel, E. M. Stein and S. Wainger, *Differentiation in lacunary directions*, Proc. Nat. Acad. Sci. U.S.A. 75 (1978), 1060–1062.
34. M. M. Rao and Z. D. Ren, *Theory of Orlicz spaces*, Marcel Dekker, New York, 1991.
35. P. Sjögren and P. Sjölin, *Littlewood-Paley decompositions and Fourier multipliers with singularities on certain sets*, Ann. Inst. Fourier 31 (1981), 157–175.
36. E. M. Stein, *Harmonic Analysis: Real-variable methods, orthogonality and oscillatory integrals*, Princeton Mathematical Series, 43. Monographs in Harmonic Analysis, III. Princeton University Press, Princeton, NJ, 1993.
37. E. M. Stein, *Maximal functions: spherical means*, Proc. Nat. Acad. Sci. U.S.A. 73 (1976), 2174-2175.
38. E. M. Stein and J.-O. Stromberg, *Behavior of maximal functions in R^n for large n*, Ark. Mat. 21 (1983), 259–269.
39. J.-O. Stromberg, *Maximal functions associated to rectangles with uniformly distributed directions*, Ann. Math. 107 (1978), 399–402.
40. N. Wiener, *The ergodic theorem*, Duke Math. J. **5** (1939), 1–18.
41. T. Wolff, *An improved bound for Kakeya type maximal functions*. Rev. Mat. Iberoamericana 11 (1995), no. 3, 651–674.
42. T. Wolff, *Recent work connected with the Kakeya problem*, Prospects in mathematics (Princeton, NJ, 1996), 129–162, Amer. Math. Soc., Providence, RI, 1999.
43. T. Wolff, *Lectures on harmonic analysis*, Edited by I. Łaba and C. Shubin. University Lecture Series, 29. American Mathematical Society, Providence, RI, 2003.
44. A. Zygmund, *Trigonometric series: Vols. I, II*, second edition, Cambridge University Press, London-New York, 1968. (Current edition in Cambridge Mathematical Library, Cambridge University Press, Cambridge, 2002.)

LINEAR DYNAMICS

GILLES GODEFROY
Institut de Mathématiques de Jussieu
Université Paris 6
4,Place Jussieu, 75252 Paris Cedex 05
E-mail: godefroy@math.jussieu.fr

This survey follows the lectures I delivered in Granada during the 2nd Course of Mathematical Analysis in Andalucía in September 2004. I am glad to thank the organizers of this meeting, and in particular Professor M. V. Velasco Collado, for the excellent working conditions and pleasant atmosphere of this international gathering.

In this note, I follow quite closely the talks I actually gave at the meeting. This means that this work is by no means an updated or comprehensive report on hypercyclic operators - a very active field of research with many contributors. I focused instead on some important new results due to Sophie Grivaux and Frédéric Bayart ([5], [6], [7], [23], [24], [25]). Even so, my account on their recent work is far from comprehensive. What I tried to do is to present some theorems and their proofs, and to avoid technicalities each time it was possible. So these notes should be seen as an invitation to further reading, an introduction to the original articles, and a preparation for the challenging questions which remain open.

We refer to [26] for a very useful and authoritative survey of the theory as it was in 1999, and [27] for an updating of this survey.

1. HYPERCYCLIC OPERATORS

Linear continuous operators (in short, operators) are relevant to every single domain of modern mathematics, and it is critically important to understand their structure. Such analysis is frequently done through spectral theory, which relies in part on the research of eigenvectors or more generally of closed invariant subspaces. This generality is needed in the infinite-dimensional setting, where very simple examples (such as the operator M defined on $L^2(\mathbb{T})$ by $M(f)(e^{i\theta}) = e^{i\theta} f(e^{i\theta})$) show that even in complex spaces there might be no eigenvector. In the Banach space setting, it might even be so that there is no non-trivial closed invariant subspace ([20]). It is an important open problem to know whether this can happen for an operator on the separable complex Hilbert space, or even on a complex reflexive space.

Finding a non-trivial closed invariant subspace for the operator $T \in L(X)$ amounts to finding a vector $x \in X$ whose orbit $O_T(x)$ under T, namely

$$O_T(x) = \{T^n(x);\ n \geq 0\}$$

does *not* span a dense linear subspace of X. It is legitimate to wonder whether a completely different behaviour may take place. In other words, can it be so that the orbit $O_T(x)$ of some vector x is actually *dense* in X? This cannot happen in the finite-dimensional case: pick an eigenvector x^* of T^* such as $T^*(x^*) = \lambda x^*$; for every $x \in X$ one has $x^*(T^n(x)) = \lambda^n x^*(x)$, thus the sequence $x^*(T^n(x))$ cannot be dense in the complex plane and in particular the orbit $O_T(x)$ is not dense. More generally, a compact operator K on a Banach space X has no dense orbit: indeed if the spectral radius $\rho(K) > 0$ then $\rho(K^*) = \rho(K) > 0$, and then the compact operator K^* has an eigenvector and the previous argument applies. On the other hand, if $\rho(K) = 0$ then the spectral radius formula shows that the sequence $(\|K^n\|)_{n\geq 0}$ is bounded and thus no orbit can be dense.

However, the existence of dense orbits is by no means an exceptional phenomenon, and we shall see below that a wealth of natural operators satisfy this condition. It is now time to give it a name:

Definition 1.1: *Let X be a Fréchet space, and let $L(X)$ be the space of continuous linear maps from X to X. An operator $T \in L(X)$ is called an hypercyclic operator if there exists $x \in X$ whose orbit $O_T(x)$ is dense in X. Such a vector x is called T-hypercyclic. We denote by $Hyp(T)$ the set of all T-hypercyclic vectors.*

We recall that a Fréchet space is a locally convex metrizable complete vector space. In particular, Banach spaces are of course Fréchet spaces. As we shall see, hypercyclicity can take place in the frame of Banach spaces. It can be said very roughly that integration operators, or forward shifts, are not hypercyclic, while their conjugate operators (differentiation operators, backward shifts) often are. We will meet in these notes many examples which illustrate this thumb rule.

2. BAIRE CATEGORY TECHNIQUES

The following simple lemma shows that hypercyclicity can be handled through the inverse (or direct) image of open sets.

Lemma 2.1: *Let X be a separable Fréchet space, and $T \in L(X)$ be an operator on X. Then T is hypercyclic if and only if for every nonempty open set U, the set $\cup_n T^{-n}(U)$ is a dense subset of X.*

Proof: If T is hypercyclic, then for every $x \in Hyp(T)$ we clearly have $O_T(x) \subset Hyp(T)$, and thus $Hyp(T)$ is dense. Since for any U, we have $Hyp(T) \subset \cup_n T^{-n}(U)$, this latter set is dense as well.

For showing the converse implication, take $(U_k)_k$ a countable basis of the topology of X. If for all k, the set $\cup_n T^{-n}(U_k)$ is dense (and open since T is continuous), then by Baire's lemma, the set

$$\Omega = \cap_k \cup_n T^{-n}(U_k)$$

is a dense subset of X, and it is immediate that actually $\Omega = Hyp(T)$. □

Note that it also follows that T is hypercyclic if and only if $\cup T^n(U)$ is dense for every nonempty open set U. When T is invertible, this implies that T is hypercyclic if and only if T^{-1} is hypercyclic. Moreover, the proof of Lemma 2.1 shows that if $Hyp(T) \neq \emptyset$, then $Hyp(T)$ is a dense G_δ, that is, "almost every" $x \in X$ in the Baire category sense is T-hypercylic.

Our next Lemma is the so-called "Hypercyclicity Criterion", which refines a previous result due to Kitai [30].

Lemma 2.2: *Let X be a separable Fréchet space, and $T \in L(X)$ be such that there exist a strictly increasing sequence (n_k) of positive integers, two dense subsets V and W of X and a sequence (S_{n_k}) of maps $S_{n_k} : W \to X$ such that:*

1.(T^{n_k}) converges to 0 pointwise on V.

2.(S_{n_k}) converges to 0 pointwise on W.

3. For every $w \in W$, the sequence $(T^{n_k} S_{n_k} w)$ converges to w.

Then T is hypercyclic.

Proof: By lemma 2.1 we have to show that if U and U' are two nonempty open sets then $\cup_n T^{-n}(U) \cap U' \neq \emptyset$. Pick $w \in W \cap U$ and $v \in V \cap U'$. We have

$$T^{n_k}(v + S_{n_k} w) = T^{n_k} v + T^{n_k} S_{n_k} w$$

and for k large enough, we have $v + S_{n_k} w \in U'$ and $T^{n_k} v + T^{n_k} S_{n_k} w \in U$. The result follows. □

Note that no assumption at all is made on the maps (S_{n_k}), which do not have to be linear nor continuous.

No example is known of an hypercyclic operator which would fail the conditions of the Hypercyclicity Criterion. The problem goes back to [31]. This may look at first sight as an artificial question. But we shall see in Section 5 why this is somehow the main remaining problem of the theory.

The next Lemma, which is sometimes called the Godefroy-Shapiro Criterion, is a special case of the Hypercyclicity Criterion.

Lemma 2.3: *Let T be an operator on a separable Fréchet space X. If the union of the eigenspaces $Ker(T - \lambda I)$ over $|\lambda| > 1$ spans a dense subspace X^+, and the union of the eigenspaces $Ker(T - \lambda I)$ over $|\lambda| < 1$ spans a dense subspace X^-, then T is hypercyclic.*

Proof: We note that T is an invertible linear map from X^+ onto itself. In the notation of Lemma 2.2 we take $V = X^-$, $W = X^+$ and $S = T^{-1} : W \to W$ the inverse of T on X^+. The conditions of Lemma 2.2 are clearly satisfied. □

So, when the point spectrum of T has "many points" on both sides of the unit circle, then T is hypercyclic. We shall see in Section 3 the deeper result that it suffices that the point spectrum meets the unit circle on a "large set" for obtaining an even stronger conclusion.

Applications 2.4: - If $B : l^2(\mathbb{N}) \to l^2(\mathbb{N})$ is the backward shift such that $B(e_0) = 0$ and $B(e_n) = e_{n-1}$, and if $|\alpha| > 1$, then $T = \alpha B$ is hypercyclic [37]. The point spectrum is the open disc of radius $|\alpha|$ and Lemma 2.3 is easily seen to apply.

- Let $X = \mathcal{H}(\mathbb{C})$ be the Fréchet space of entire functions on $\mathbb{C}$. Let $T \in L(X)$ an operator such that $T\tau_z = \tau_z T$ for all $z \in \mathbb{C}$, where $\tau_z(f)(u) = f(z + u)$. Such an operator T is sometimes called a pseudo-differential operator, and here is why. Let δ_0 be the Dirac measure at 0, such that $\delta_0(f) = f(0)$, and let $\psi = T^*(\delta_0)$. For all $f \in X$ we have

$$Tf(z) =< \delta_0, \tau_z Tf >=< \delta_0, T\tau_z f >= \psi(\tau_z f).$$

On the other hand, Taylor's formula reads

$$\tau_z f = \sum_{k=0}^{\infty} \frac{f^{(k)}(z)}{k!} p_k$$

where $p_k(u) = u^k$ and this series converges in X. Therefore

$$Tf(z) = \psi(\tau_z f) = \sum_{k=0}^{\infty} \frac{f^{(k)}(z)}{k!} \psi(p_k) = \sum_{k=0}^{\infty} \frac{\psi(p_k)}{k!} D^k(f)(z)$$

where $D(g) = g'$ is the derivation operator, and since this is true for every z and every f, we have shown that $T = F(D)$, where

$$F(u) = \sum_{k=0}^{\infty} \frac{\psi(p_k)}{k!} u^k$$

is an entire function.

So any operator T which commutes with all translations can be represented as $T = F(D)$ with F an entire function, and conversely every such operator commutes with all translations since D does, provided of course that the operator $F(D)$ is well-defined (that is, F is of exponential type).

Assume now that T commutes with the translations and is not proportional to the identity operator. We have $T = F(D)$ where F is not a constant function. For any $\lambda \in \mathbb{C}$, let $e_\lambda(z) = e^{\lambda z}$. It is clear that $T(e_\lambda) = F(\lambda)e_\lambda$. Since F is a non-constant entire function, the open sets $U^- = F^{-1}(\{|z| < 1\})$ and $U^+ = F^{-1}(\{|z| > 1\})$ are both non empty. For applying Lemma 2.3 with $X^- = span\{e_\lambda;\ \lambda \in U^-\}$ and $X^+ = span\{e_\lambda;\ \lambda \in U^+\}$, it suffices to show that if U is a nonempty open set of $\mathbb{C}$, the space $E_U = span\{e_\lambda :\ \lambda \in U\}$ is dense in X.

We show this by a Hahn-Banach argument. We observe that for every continuous linear form ϕ on X, the function $G(\lambda) = \phi(e_\lambda)$ is an entire function, and an easy induction shows that $G^{(k)}(0) = \phi(p_k)$. If ϕ vanishes on E_U, then G vanishes on U and therefore everywhere. Hence $G^k(0) = \phi(p_k) = 0$ for every $k \geq 0$, and thus $\phi = 0$ since the polynomials are dense in X.

So Lemma 2.3 applies and we have shown the following result from [22]:

Theorem 2.5: *Every operator on $\mathcal{H}(\mathbb{C})$ which commutes with translations and is not proportional to the identity is hypercyclic.*

This result extends the first hypercyclicity theorems, due to Birkhoff [13] (for the translation operator τ_1) and Mac Lane [33] (for the derivation operator D). Easy modifications of the proof show [22] that the above theorem extends to $\mathcal{H}(\mathbb{C}^n)$, and also to the real frame of differential operators on $\mathcal{C}^\infty(\mathbb{R}^n)$. Note that such a result, which provides a (very large) cone of operators which are hypercyclic, cannot take place with Banach spaces since of course every hypercyclic operator on a Banach space has norm greater than one.

3. THE ROLE OF THE UNIMODULAR POINT SPECTRUM

Most of the techniques from this section are rather quantitative, while so far qualitative arguments have been used. Moreover, residual sets in the Baire category sense now give way to sets of full measure, with respect to appropriate measures on the space where the operators are defined.

This section relies on the recent and deep work of F. Bayart and S. Grivaux. This first definition comes from [7].

Definition 3.1: *An operator T on a complex Banach space X is said to have a perfectly spanning set of eigenvectors associated to unimodular eigenvalues if there exists a probability measure σ on the unit circle $\mathbb{T}$ which is continuous (that is, $\sigma(\{\lambda\}) = 0$ for every $\lambda \in \mathbb{T}$) and such that for every Borel subset A of $\mathbb{T}$ with $\sigma(A) = 1$, the space $span(\cup\{Ker(T - \lambda I) : \lambda \in A\})$ is dense in X. In short, we will say that T has a perfectly spanning set (with respect to σ).*

Example 3.2: if B is the backward shift on the space $l_p(\mathbb{N})$, $1 \leq p < \infty$ and $|\omega| > 1$, the operator $T = \omega B$ has a perfectly spanning set with respect to the Haar measure of the circle group.

This notion allows to establish hypercyclicity through the unimodular point spectrum. This is somehow unexpected since the corresponding eigenvectors have very simple (and bounded) orbits. However if there are enough of them, they provide by some kind of resonance vectors with dense orbits.

Theorem 3.3: *Every operator with a perfectly spanning set is hypercyclic.*

Proof: We outline the proof in the case when the continuous measure σ is actually absolutely continuous with respect to the Haar measure and when the spectrum is simple. There is a measurable vector field $E : \mathbb{T} \to X$ such that $E(\lambda) \neq 0$ if and only if λ is an eigenvalue and $TE(\lambda) = \lambda E(\lambda)$ in this case, and $\|E(\lambda)\| \leq 1$ everywhere.

Let us define $K : L^2(\mathbb{T}, \sigma) \to X$ by

$$K(g) = \int_{\mathbb{T}} g(\lambda) E(\lambda) d\sigma(\lambda).$$

We define also a multiplication operator V on $L^2(\mathbf{T}, \sigma)$ by $Vf(\lambda) = \lambda f(\lambda)$. In this notation we clearly have the operator equation

$$TK = KV \tag{F}$$

Let $X_0 = K(L^2(\mathbb{T}, \sigma))$. If $x^* \in X_0^{\perp}$, it follows that $< x^*, E(\lambda) >= 0$ for σ-almost every λ and the perfectly spanning set condition shows that $x^* = 0$. Therefore X_0 is dense.

If $x = K(g) \in X_0$, we have

$$T^n(x) = \int_{\mathbb{T}} \lambda^n g(\lambda) E(\lambda) d\sigma(\lambda).$$

Since σ is absolutely continuous, the Fourier coefficients of the X-valued measure $gEd\sigma$ tend to zero at infinity. Hence T^n converges to zero pointwise on the dense subspace X_0. We now define maps $S_n : X_0 \to X_0$ as follows: for any $x \in X_0$, pick g such that $x = K(g)$, and define

$$S_n(x) = \int_{\mathbf{T}} \lambda^{-n} g(\lambda) E(\lambda) d\sigma(\lambda).$$

We have $T^n S_n = Id_{X_0}$ for all n, and the same argument as before shows that S_n converges to zero pointwise on X_0. We can therefore apply the Hypercyclicity Criterion (Lemma 2.2) to conclude that T is hypercyclic. □

Remark 3.4: When the measure σ is absolutely continuous, the hypercyclicity criterion is applied in the above proof with the whole sequence of integers and not only with a subsequence. It follows that under the assumptions of Theorem 3.3 the operator T is *topologically mixing*, that is, if U and V are nonempty open sets then $T^n(U) \cap V \neq \emptyset$ for n large enough. Note that this is also true for operators which satisfy the assumptions of Lemma 2.3.

When σ is simply continuous, it is not so in general that its Fourier coefficients tend to zero, but Wiener's theorem states that the Cesaro means of the Fourier coefficients converge to zero, and this suffices to use the Hypercyclicity Criterion.

The equation $TK = KV$ from the proof of Theorem 3.3 is called Flytzanis equation since it has been introduced in Flytzanis' seminal article [21]. Analysing the proof of Theorem 3.3 (following [18]) yields to the following observation.

Proposition 3.5: *Let X be a Banach space, and $T \in L(X)$ be such that there is a unitary operator $V \in L(H)$ such that $V^n(x)$ tends weakly to zero for every $x \in H$, and a compact operator $K : H \to X$ with dense range such that $TK = KV$. Then T is hypercyclic.*

Proof: Let $X_0 = K(H)$. If $y = K(x) \in X_0$, we have $T^n(y) = KV^n(x)$ and since K is compact it follows that $\lim \|T^n(y)\| = 0$. We define the map $S_n : X_0 \to X$ by $S_n(y) = KV^{-n}(x)$, where x such that $K(x) = y$ has been chosen once and for all. Since V is unitary, $V^{-1} = V^*$ and the same argument as before shows that $\lim \|S_n(y)\| = 0$ for all $y \in X_0$. Finally we clearly have $T^n S_n = Id_{X_0}$. It suffices therefore to apply Lemma 2.2. □

Following [7] we now use Flytzanis' equation (F), in the Hilbertian case, to construct a T-invariant Gaussian measure.

Lemma 3.6: *Let $T \in L(H)$ an operator acting on the separable complex Hilbert space H, which has a perfectly spanning set. Then there exists a non-degenerate Gaussian T-invariant measure m on H.*

Proof: In the notation of the proof of Theorem 3.3, T satisfies Flytzanis equation

$$TK = KV \tag{F}$$

Note that in the equation (F), K is a Hilbert-Schmidt compact operator from $L^2(\mathbb{T}, \sigma)$ to H, while V is a unitary operator on $L^2(\mathbb{T}, \sigma)$. We set $S = KK^*$. The operator S is a self-adjoint positive trace class operator, and S is injective since K has dense range. Hence there exists a non-degenerate Gaussian measure m on H such that S is the covariance operator of m (see [8], section 6.2): that is,

$$< Sx, y >= \int_H < x, z > \overline{< y, z >} dm(z)$$

The measure $m \circ T^{-1}$ is also a Gaussian measure and its covariance operator S' is given by

$$< S'x, y >= \int_H < x, z > \overline{< y, z >} d(m \circ T^{-1})(z) = \int_H < T^*x, z > \overline{< T^*y, z >} dm(z)$$

hence

$$< S'x, y >=< ST^*x, T^*y >=< TST^*x, y >$$

and thus $S' = TST^*$. Using (F) again, we now have

$$TST^* = TKK^*T^* = KVV^*K^* = KK^* = S$$

since V is unitary, and this shows that m is T-invariant. □

We recall that a measure-preserving transformation T is said to be weak-mixing if for all measurable sets A and B, one has

$$\lim_{N \to \infty} \frac{1}{N} \sum_{k=0}^{N-1} |m(T^{-k}(A) \cap B) - m(A)m(B)| = 0$$

and T is called strong-mixing if moreover

$$\lim_{N \to \infty} |m(T^{-k}(A) \cap B) - m(A)m(B)| = 0.$$

Weak-mixing and strong-mixing transformations are in particular ergodic.

We now state an important result from [7]:

Theorem 3.7: *If $T \in L(H)$ has a perfectly spanning set with respect to σ, then there exists a non-degenerate T-invariant Gaussian measure m on H such that T is weak-mixing with respect to m. If moreover σ is absolutely continuous with respect to the Haar measure, then T is strong-mixing.*

Proof: We outline the proof in the absolutely continuous case. Lemma 3.6 provides a T-invariant Gaussian measure m, and T induces an isometry U_T on $L^2(H, m)$ defined by $U_T(f) = f \circ T$. Showing that T is strong-mixing amounts to show that for all f and g in $L^2(H, m)$,

$$\lim_{n\to\infty} < U_T^n f, g >=< f, 1 > \overline{< g, 1 >}$$

We show it when $f = f_x$ and $g = f_y$, with $f_x(z) =< z, x >$. Since these functions have mean zero, we have to show that

$$\lim_{n\to\infty} < U_T^n f_x, f_y >= 0. \tag{1}$$

We compute

$$< U_T^n f_x, f_y >= \int_H < T^k z, x > \overline{< z, y >} dm(z) =< T^k Sy, x >$$

where S is the covariance operator from Lemma 3.6. But since by (F), we have $TK = KV$, it follows that $T^k K = KV^k$ for all $k \geq 1$, and since $S = KK^*$, it follows that $T^k S = KV^k K^*$. This finally shows that

$$< U_T^n f_x, f_y >=< V^k K^* y, K^* x >$$

We let now $h = K^*y.K^*x$, where the dot denotes the pointwise product of functions on $\mathbb{T}$. The function h belongs to $L^1(\mathbb{T}, \sigma)$, and since σ is absolutely continuous, it follows that the Fourier coefficients of h tend to zero at infinity. But we clearly have

$$< V^k K^* y, K^* x >= \hat{h}(-k)$$

and thus (1) is shown.

It remains to prove (1) for arbitrary functions in $L^2(H, m)$. This requests the use of Fock spaces, and for this we refer to [7].

The case when σ is simply continuous follows the same lines, except that the convergence to zero of the Fourier coefficients is replaced by Wiener's theorem, which yields to the weak-mixing property. □

Following [7], we now introduce a quantitative form of hypercyclicity.

Definition 3.8: *Let X be a Fréchet space. An operator $T \in L(X)$ is called frequently hypercyclic if there exists a vector $x \in X$ such that for every nonempty open set U, the set $\{n \in \mathbb{N};\ T^n(x) \in U\}$ has positive lower density. Such a vector x is said to be T-frequently hypercyclic.*

In other words, not only the orbit $O_T(x)$ is dense but moreover it meets every open set frequently enough. The next result, still from [7], provides such operators. It can be seen as a quantitative version of Proposition 3.5.

Theorem 3.9: *If $T \in L(H)$ has a perfectly spanning set, then T is frequently hypercylic.*

Proof: By Theorem 3.7, there exists a non-degenerate Gaussian measure m on H with respect to which T is ergodic. By Birkhoff's ergodic theorem, for m-almost every $x \in H$ we have

$$\lim_{N\to\infty} \frac{1}{N}\sharp\{k \leq N;\ T^k x \in U\} = m(U) \tag{2}$$

If (U_k) is a countable basis of the topology of H, it follows that for m-almost every $x \in H$, the equation (2) holds for every U_k. Since m is non-degenerate, $m(U_k) > 0$ for all k. Hence m-almost every $x \in H$ is a frequently hypercyclic vector. □

Remarks 3.10: The proof of Lemma 2.1 shows immediately that countably many hypercylic operators have a common hypercyclic vector, and that if T is hypercyclic and invertible then T^{-1} is hypercyclic. It is not known whether one may replace "hypercyclic" by "frequently hypercyclic" in the above statements. It is not even known whether two frequently hypercyclic operators have a common frequently hypercyclic vector, even if these operators both have a perfectly spanning set. Indeed it can happen that the corresponding Gaussian measures are mutually singular. Note that the set of frequently hypercyclic vectors of a frequently hypercyclic operator is in general a meager set [15].

Examples 3.11: - If B is the backward shift on $l_2(\mathbb{N})$ and $|\omega| > 1$, then $T = \omega B$ has a perfectly spanning set and is thus frequently hypercyclic. By using a different technique, based on a Frequent Hypercyclicity Criterion, it is shown in [7] that the conclusion extends to $l_p(\mathbb{N})$, $1 \leq p < \infty$, and this also follows from [25] by analyticity of the field of unimodular eigenvectors.

- Let Ω be a connected open subset of $\mathbb{C}$, and let H be a Hilbert space of analytic functions on Ω such that the point evaluations δ_z $(z \in \Omega)$ are continuous. Let ϕ be a multiplier, that is, $M_\phi(f) = \phi f$ defines an operator from H to H. It was shown in [22] that the adjoint M_ϕ^* is hypercyclic whenever ϕ is non-constant and $\phi(\Omega)$ intersects the unit circle. Under this assumption, M_ϕ^* has a perfectly spanning set (with respect to the normalized Lebesgue measure on some subarc of $\mathbb{T}$) and thus it is frequently hypercyclic.

- Let φ be an automorphism of the unit disc $\mathbb{D}$ and C_φ be the composition operator defined on the Hardy space $H^2(\mathbb{D})$ by $C_\varphi(f) = f \circ \varphi$. It is shown in [7] that parabolic and hyperbolic composition operators have a perfectly spanning set, and are therefore frequently hypercyclic. We refer to [6] for the existence of common hypercyclic vectors for some uncountable families of hypercyclic composition operators, and to [1] for the similar result for all hypercyclic multiples of the backward shift.

- Birkhoff's operator τ_1 and Mac Lane's operator D on $\mathcal{H}(\mathbb{C})$ are both frequently hypercyclic, as can be seen from the Frequent Hypercyclicity Criterion. More generally, it has been shown in [14] that all the operators considered in Theorem 2.5, as well as their counterparts on $\mathcal{H}(\mathbb{C}^n)$, are frequently hypercyclic.

- Theorems 3.7 and 3.9 concern operators on the Hilbert space. Extending them to Banach spaces is not straightforward, since for instance no characterization of covariance operators of Gaussian measures is available in the Banach space setting. However it is shown in [25] that Theorems 3.7 and 3.9 extend when H is replaced by a Banach space X of type 2. With the notation of the proof of Theorem 3.3, if $T \in L(X)$ where X is an arbitrary Banach space and the map E is α-Hölder with $\alpha > 1/2$, then there is a T-invariant Gaussian measure on X with respect to which T is strongly mixing [25].

- The weighted backward shift T on $l_2(\mathbb{N})$ with weights $\omega_n = \sqrt{1 + 1/n}$, which can also be seen as the backward shift on the Bergman space A_2, is a hypercyclic operator which is not frequently hypercyclic [7]. This operator T satisfies Flytzanis' equation with a compact operator (the canonical injection from A_2 to $L^2(\mathbb{T})$) which is not Hilbert-Schmidt. Proposition 3.5 proves hypercyclicity, but falls short to show the quantitative version (frequent hypercyclicity) for which the Hilbert-Schmidt condition is needed.

- There is [25] a frequently hypercyclic operator T on $c_0(\mathbb{N})$ (actually, a particular weighted backward shift) which does not have any unimodular eigenvalue, is not topologically mixing, and such that no Gaussian measure on $c_0(\mathbb{N})$ is T-invariant.

4. OPERATORS WITH PRESCRIBED ORBIT

In this short section, we follow [23] where existence of special operators on arbitrary Banach spaces and normed spaces is shown through hypercyclicity. We first recall a result shown independently by Bourdon [16] and Herrero [29] in the complex case, and Bes [11] in the real case.

Theorem 4.1: *Let X be a real or complex Banach space, and let T be an hypercyclic operator on X. If $x \in X$ is T-hypercyclic and $P \in \mathbb{K}[\xi]$ is a non zero polynomial, then $y = P(T)(x)$ is T-hypercyclic. In particular, for* *every hypercyclic operator T there exists a dense linear subspace M of X such that every $y \in M\backslash\{0\}$ is T-hypercyclic.*

Proof: we prove it only in the complex case. If T is hypercyclic, then for every non zero polynomial P, the operator $P(T)$ has dense range. Indeed if not, there is $x_0^* \in X^*$ such that $P(T^*)(x_0^*) = 0$. It follows that x_0^* is contained in a finite-dimensional T^*-invariant subspace, and thus T^* has an eigenvector x^* such that $T^*(x^*) = \lambda x^*$. For every $x \in X$ one has $x^*(T^n(x)) = \overline{\lambda}^n x^*(x)$, thus the sequence $x^*(T^n(x))$ cannot be dense in the complex plane and in particular the orbit $O_T(x)$ is not dense, contradicting hypercyclicity.

Now if x is T-hypercyclic and $P \neq 0$, we have

$$O_T(P(T)x) = P(T)(O_T(x)) \tag{3}$$

since T and $P(T)$ commute. Since $P(T)$ has dense range and $O_T(x)$ is dense, $P(T)(O_T(x))$ is dense as well and (3) shows that $P(T)(x)$ is T-hypercyclic. □

Note that Theorem 4.1 works as well, with practically the same proof, for frequently hypercyclic operators.

Theorem 4.1 is interesting in itself, but it also yields quite unexpected consequences, shown in [23]. We first have:

Theorem 4.2: *Let X be an arbitrary real or complex infinite-dimensional separable Banach space. Let $V = \{v_n;\ n \geq 1\}$ be a dense linearly independent sequence. Then there exists an operator $T \in L(X)$ such that $O_T(v_1) = V$.*

This result shows in particular that every dense linearly independent sequence is an orbit, even if the space X is a pathological Banach space with "few" operators.

Sketch of proof: by a result due to Ansari [3] and Bernal-Gonzalez [9], there is a hypercyclic operator T_0 on X. The construction of this operator relies on a technique due to Salas [38] and $T_0 = I + K$ where K is a nuclear operator (with arbitrarily small norm). The set $Hyp(T_0)$ is dense in X. A direct construction ([23, Lemma 2.1 and Remark 2.4) provides an isomorphism $J \in L(X)$ such that $J(v_1) = x \in Hyp(T_0)$ and $J(V) = O_{T_0}(x)$. It is clear that the operator $T = J^{-1}T_0J$ works. □

Combining the above two results, we obtain a very strong negative answer to the invariant subspace problem for normed spaces of countable dimension [23].

Theorem 4.3: *Let M be a normed space of countable algebraic dimension. Then there exists a bounded linear operator T on M such that every non zero vector of M is T-hypercyclic. In other words, T has no non-trivial invariant closed subset.*

Proof: The space M has a dense algebraic basis. Indeed if (V_n) is a basis of the topology of M, we easily construct by induction a linearly independent family (v_n) such that $v_n \in V_n$ for all n. This can be done since the open sets V_n are not contained in a finite dimensional subspace. The family (v_n) is dense by construction, and it is contained in a basis of M since it is linearly independent. We relabel this dense basis as $(e_n)_{n\geq 0}$. By Theorem 4.2 applied to the completion X of M, there is a bounded linear operator T such that $O_T(e_o) = (e_n)_{n\geq 0}$. Since $(e_n)_{n\geq 0}$ is a basis of M, it follows that

$$M = \{P(T)(e_0) : \ P \in \mathbb{K}[\xi]\} \tag{4}$$

The vector (e_0) is T-hypercyclic and thus by Theorem 4.1, for every non zero $P \in \mathbb{K}[\xi]\}$ the vector $y = P(T)(e_0)$ is T-hypercyclic as well. But then (4) shows that every $y \in M\backslash\{0\}$ has a dense orbit under T. This clearly means that T has no non-trivial closed invariant subset. □

Theorem 4.3 can be formulated as follows: if $\| \ . \ \|$ is any norm on the space $\mathbb{K}[\xi]$ of polynomials, there is a $\| \ . \ \|$-continuous linear operator on $\mathbb{K}[\xi]$ with no non-trivial closed invariant subset.

It is considerably more difficult to construct such an operator on a Banach space. This has been done by Read ([36]) on the Banach space $l_1(\mathbb{N})$. It is not known whether such an operator could exist on a reflexive Banach space.

Remark 4.4: In the notation of Theorem 4.3, the algebra $\mathbb{K}[\xi]$ normed by $\|P\| = \|P[T](e_0)\|_M$ is a commutative algebra without non-trivial closed ideals. Note however that the inequality $\|PQ\| \leq \|P\|.\|Q\|$ is not satisfied.

5. THE MAIN PROBLEM

The concept of hypercyclicity can be strengthened in various ways. We have already met the concept of frequent hypercyclicity (Definition 3.8).

Let us now list other notions. The letter T denotes as before an operator on a Fréchet space X.

- T is (topologically) mixing if for every nonempty subsets U and V of X, there exists N such that $T^n(U) \cap V \neq \emptyset$ for every $n \geq N$.

- T is chaotic if it is hypercyclic and has a dense set of periodic points. In the original definition due to Devaney [19], a "sensitivity to initial conditions" was also assumed, but this last condition happens to be redundant [4].

- T is hereditarily hypercyclic with respect to the (strictly increasing) sequence (n_k) if for every subsequence (n_{k_j}) of (n_k), there exists $x \in X$ such that the sequence $(T^{(n_{k_j})}(x))_j$ is dense. In other words, every subsequence $(T^{(n_{k_j})})_j$ is hypercyclic.

- T is doubly hypercyclic if there exist a pair $(x, y) \in X^2$ such that for every $(x', y') \in X^2$ and every $\epsilon > 0$, there is $n \in \mathbb{N}$ such that $\|T^n(x) - x'\| < \epsilon$ and $\|T^n(y) - y'\| < \epsilon$.

Note of course that the same integer n must work for both x' and y'. In other words, T is doubly hypercyclic if $T \oplus T$ is hypercyclic on X^2. This terminology is not really used and we will see why in a moment. Let us now formulate the

MAIN PROBLEM: Let T be a hypercyclic operator on a separable Fréchet space X. Is $T \oplus T$ hypercyclic on X^2?

Let us make clear that the Main Problem is open as well when X is a Banach space, or even the separable Hilbert space.

What makes this problem particularly fascinating is that it turns out to be equivalent to several problems which have been independently formulated. This is due to the following theorem, which results from [10], [12], [35], [24]:

Theorem 5.1: *Let X be a separable Banach space, and $T \in L(X)$ be a hypercyclic operator on X. The following assertions are equivalent:*

(i) $T \oplus T$ is hypercyclic.

(ii) T satisfies the hypercyclicity criterion (see Lemma 2.2).

(iii) $T \oplus T$ is cyclic.

(iv) T is hereditarily hypercyclic (with respect to some strictly increasing sequence).

(v) For every increasing sequence (n_k) such that the sequence of differences $(n_{k+1} - n_k)$ is uniformly bounded, the sequence (T^{n_k}) is hypercyclic.

(vi) For every increasing sequence (n_k) such that $(n_{k+1} - n_k) \leq 2$ for

every k, the sequence (T^{n_k}) is hypercyclic.

When these equivalent conditions are satisfied, we say that T satisfies the Hypercyclicity Criterion (HC). So, the Main Problem amounts to asking whether every hypercyclic operator T satisfies (HC).

It is probably appropriate to state that every operator which is hypercyclic "for some good reason" satisfies (HC). Indeed each of the following conditions on T implies that T satisfies (HC):

- T is mixing [12].

- T is chaotic [12], or more generally T is a hypercyclic operator which has a dense set of vectors whose orbit is bounded [24].

- T is frequently hypercyclic [28].

Note that each of these conditions (mixing, chaotic, frequently hypercyclic) is actually strictly stronger than satisfying (HC). The properties "mixing" and "frequently hypercyclic" are independent [25], and "frequently hypercyclic" does not imply "chaotic". The converse seems to be an open question. I have been informed by S. Grivaux that "chaotic" and "mixing" are independent as well (for operators on the Hilbert space).

- if T is a hypercyclic operator such that

$$\bigcup_{P\in\mathbb{K}[\xi]\setminus\{0\}} Ker P[T]$$

is dense in X, then T satisfies (HC) [24]. This applies in particular to upper-triangular hypercyclic operators, such as the numerous examples of the form $I+B$, where B is a weighted backward shift. It applies as well to hypercyclic operators such that the linear span of the eigenvectors is dense.

It seems therefore that any hypercyclic operator which has sufficiently many points with a "regular" orbit satisfies (HC). This is not the case of Read's operator [36] for which *every* non-zero vector is hypercyclic. And indeed it is not known, to the best of my knowledge, whether Read's operator satisfies (HC). If we think of an operator (say, on the Hilbert space) as an infinite matrix, we can somehow handle it when it is "concentrated around the diagonal", but if it has large coefficients "in the corners" (even if we try to choose the basis properly), then what can we do?

We finally illustrate the Main Problem with some related results.

Remarks 5.2: Even when T satisfies (HC), one has to pick special hypercyclic vectors for establishing conditions (v) or (vi). This has been shown in [34]. Indeed, let $x \in X$ be a hypercyclic vector. Of course $T(x) \neq x$ and there is an open neighbourhood U of x such that $T(U) \cap U = \emptyset$.

We now enumerate the set $\mathcal{S} = \{n : T^n(x) \notin U\}$. Clearly, if $n \notin \mathcal{S}$ then $(n+1) \in \mathcal{S}$ and thus the sequence $\mathcal{S}$ satisfies the condition (vi) of Theorem 5.1. On the other hand, x is not $\mathcal{S}$-hypercyclic, that is, the set $\{T^n(x) : n \in \mathcal{S}\}$ is not dense in X. Indeed this set does not meet U.

However, when the sequence $\mathcal{S}$ is somewhat "regular", then every hypercyclic vector is $\mathcal{S}$-hypercyclic. Indeed, it has been shown by Ansari [2] that if T is hypercyclic, then for any integer N, the operator T^N is hypercyclic and moreover $Hyp(T) = Hyp(T^N)$. This follows from a later result by Bourdon and Feldman [17] which states: if R is an operator, $y \in X$ and the closure of $O_R(y)$ has a non-empty interior, then $O_R(y)$ is dense. Indeed, if $x \in Hyp(T)$, by Baire's lemma the closure of one of the sets $E_j = \{T^{(j+kN)}(x);\ k \in \mathbb{N}\}$, where $0 \leq j \leq (N-1)$, has a non-empty interior and then we can apply the Bourdon-Feldman theorem to $R = T^N$ and $y = T^j(x)$.

Hence the arithmetic progressions $\mathcal{S}$ behave well with respect to $\mathcal{S}$-hypercyclicity. The question to know which subsets of $\mathbb{N}$ behave well in this sense seems to be widely open, but [34] shows anyway than positive density does not suffice. Let us also mention that "rotations" of hypercyclic operators are hypercyclic as well: more precisely, if T is hypercyclic and λ is any complex number of modulus 1, then λT is hypercyclic and moreover $Hyp(T) = Hyp(\lambda T)$ [32].

An analogue for frequent hypercyclicity of the Bourdon-Feldman theorem has been shown by Grosse-Erdmann and Peris [28]: if $x \in X$ and there exists a non-empty open subset U such that $\{n;\ T^n(x) \in V\}$ has positive lower density for every non-empty open subset V of U, then x is frequently hypercyclic for T. This provides an alternative proof (and an improvement) of a result from [7] asserting that T^N is frequently hypercyclic when T is, with the same frequently hypercyclic vectors. Let us mention that the Main Problem has its "frequent" counterpart: it is not known whether $T \oplus T$ is frequently hypercyclic when T is.

An analogue (and earlier) version of the hypercyclicity criterion is Kitai's criterion [30], where the sequence (n_k) (see Lemma 2.2) is replaced by the full set $\mathbb{N}$. The proof of Lemma 2.2 then shows that every operator which satisfies Kitai's criterion is topologically mixing, and the question (obviously similar to the Main Problem) occurs of the converse. This question has been answered negatively by S. Grivaux [24]: she showed in fact that every infinite-dimensional separable Banach space supports a topologically mixing operator which fails Kitai's criterion.

References

1. Abakumov, E. and Gordon, J., Common hypercyclic vectors for multiples of backward shift, *J. Funct. Anal.* **200** (2003), 494-504.
2. Ansari, S. I., Hypercyclic and cyclic vectors, *J. Funct. Anal.* **128** (1995), 374-383.
3. Ansari, S. I., Existence of hypercyclic operators on topological vector spaces, *J. Funct. Anal.* **148** (1997), 384-390.
4. Banks, J., Brooks, J., Cairns, G., Davis, G., and Stacey, P., On Devaney's definition of chaos, *Amer. Math. Monthly* **99** (1992), 332-334.
5. Bayart, F. and Grivaux, S., Hypercyclicité: le rôle du spectre ponctuel unimodulaire, *C. R. Math. Acad. Sci. Paris* **338** (2004), 703-708.
6. Bayart, F. and Grivaux, S., Hypercyclicity and unimodular point spectrum, *J. Funct. Anal.* **226** (2005), 281-300.
7. Bayart, F. and Grivaux, S., Frequently hypercyclic operators, *Trans. Amer. math. Soc.* (to appear), 2006.
8. Benyamini, Y. and Lindenstrauss, J., *Geometric nonlinear functional analysis. Vol. 1*, Amer. Math. Soc. Colloquium Publications **48**. Amer. Math. Soc., Providence, RI, 2000.
9. Bernal-González, L., On hypercyclic operators on Banach spaces, *Proc. Amer. Math. Soc.* **127** (1999), 1003-1010.
10. Bernal-González, L. and Grosse-Erdmann, K.-G., The hypercyclicity criterion for sequences of operators, Studia Math. **157** (2003), 17-32.
11. Bès, J. P., Invariant manifolds of hypercyclic vectors for the real scalar case, *Proc. Amer. Math. Soc.* **127** (1999), 1801-1804.
12. Bès, J. P. and Peris, A., Hereditarily hypercyclic operators, *J. Funct. Anal.* **167** (1999), 94-112.
13. Birkhoff, G. D., Démonstration d'un théorème élémentaire sur les fonctions entières, *C. R. A. S. Paris*, **189** (1929), 473-475.
14. Bonilla, A. and Grosse-Erdmann, K.-G., On a theorem of Godefroy and Shapiro, *Integral Equations Operator Theory* (to appear).
15. Bonilla, A. and Grosse-Erdmann, K.-G., Frequently hypercyclic operators, Preprint.
16. Bourdon, P. S., Invariant manifolds of hypercyclic vectors, *Proc. Amer. Math. Soc.* **118** (1993), 845-847.
17. Bourdon, P. S. and Feldman, N. S., Somewhere dense orbits are everywhere dense, *Indiana Univ. Math. J.* **52** (2003), 811-819.
18. Bourdon, P. S. and Shapiro, J. H., Hypercyclic operators that commute with the Bergman backward shift, *Trans. Amer. Math. Soc.* **352** (2000), 5293-5316.
19. Devaney, R. L., *An introduction to chaotic dynamical systems*, Addison-Wesley Studies in Nonlinearity. Addison-Wesley Publishing Company Advanced Book Program, Redwood City, CA, Second edition, 1989.
20. Enflo, P., On the invariant subspace problem for Banach spaces, *Acta Math.* **158** (1987), 213-313.
21. Flytzanis, E., Unimodular eigenvalues and linear chaos in Hilbert spaces,

Geom. Funct. Anal. **5** (1995), 1-13.

22. Godefroy, G. and Shapiro, J. H., Operators with dense, invariant, cyclic vector manifolds, *J. Funct. Anal.* **98** (1991), 229-269.
23. Grivaux, S., Construction of operators with prescribed behaviour, *Arch. Math. (Basel)* **81** (2003), 291-299.
24. Grivaux, S., Hypercyclic operators, mixing operators, and the bounded steps problem, *J. Operator Theory* **54** (2005), 147-168.
25. Grivaux, S., Invariant Gaussian measures for operators on Banach spaces and linear dynamics. Preprint, 2005.
26. Grosse-Erdmann, K.-G., Universal families and hypercyclic operators, *Bull. Amer. Math. Soc. (N.S.)* **36** (1999), 345-381.
27. Grosse-Erdmann, K.-G., Recent developments in hypercyclicity, *RACSAM Rev. R. Acad. Cienc. Exactas Fís. Nat. Ser. A Mat.* **97** (2003), 273-286.
28. Grosse-Erdmann, K.-G. and Peris, A., Frequently dense orbits, *C. R. Math. Acad. Sci. Paris* **341** (2005), 123-128.
29. Herrero, D. A., Limits of hypercyclic and supercyclic operators, *J. Funct. Anal.* **99** (1991), 179-190.
30. Kitai, C., *Invariant closed sets for linear operators*, PhD thesis, Univ. of Toronto, 1982.
31. León-Saavedra, F. and Montes-Rodríguez, A., Linear structure of hypercyclic operators, *J. Funct. Anal.* **148** (1997), 524-545.
32. León-Saavedra, F. and Müller, V., Rotations of hypercyclic and supercyclic operators, *Integral Equations Operator Theory* **50** (2004), 385-391.
33. MacLane, G. R., Sequences of derivatives and normal families, *J. Analyse Math.* **2** (1952), 72-87.
34. Montes-Rodríguez, A. and Salas, H. N., Supercyclic subspaces: spectral theory and weighted shifts, *Adv. Math.* **163** (2001), 74-134.
35. Peris, A. and Saldivia, L., Syndetically hypercyclic operators, *Integral Equations Operator Theory* **51** (2005), 275-281.
36. Read, C. J., A solution to the invariant subspace problem on the space l_1, *Bull London Math. Soc.* **17** (1985), 305-317.
37. Rolewicz, S., On orbits of elements, *Studia Math.*, **32** (1969), 17-22.
38. Salas, H. N., Hypercyclic weighted shifts, *Trans. Amer. Math. Soc.* **347** (1995), 993-1004.

GREEDY ALGORITHMS AND BASES FROM THE POINT OF VIEW OF BANACH SPACE THEORY

N. J. KALTON
Department of Mathematics
University of Missouri-Columbia
Columbia, MO 65211

1. Introduction

Imagine that you are presented with the problem of transmitting a photograph over a telephone connection. The restrictions on you are the physical limitations of the telephone line, or time constraints, but you have perfect knowledge of the photograph. How do you decide what parameters to transmit? In mathematical terms the photograph may be represented by a function, f, or a family of functions representing color densities. So if we convert the problem to a more abstract level we are given a function and want to select a certain finite set of parameters to send to another person so that he or she can reconstruct the function as accurately as possible.

Of course one solution is simply to send as many points of the graph of f as is practical. If f were a completely random function, this might well be the best that can be done. But of course in practice our function is not random and contains many patterns. This suggests a series expansion of the function; one might think of Fourier series, but perhaps an expansion with respect to some wavelet basis is more appropriate.

Now the function is coded by its coefficients with respect to our given basic functions. We can only send some of these coefficients. Which ones should we choose to send? Suppose we can send a thousand coefficients. Mathematically, we are used to the idea that we pick the first one thousand as our approximation. But this may not be the most efficient for our given function; perhaps all these coefficients are small and some later coefficients are large. So, perhaps we should consider sending the thousand largest coefficients.

This is the basic idea behind the concept of a greedy basis. Of course we have been quite imprecise and we would need to specify some normalization of our basis elements and some space functions within which to work. The idea has been around for some time; the author was first introduced to it at a conference on Littlewood-Paley theory in 1990 [9]. However, the formal development of a theory of greedy bases is more recent and was initiated in an important paper of Konyagin and Temlyakov [15], which we will discuss below. Subsequently the theory has been developed quite rapidly from the point of view of approximation theory by Temlyakov and others; we refer to [21] for a recent survey.

We are going to concentrate on the Banach space aspects of this theory, where rather unexpectedly the the theory of greedy bases has links to some old and classical results and also to some open problems. The idea of studying greedy bases and related greedy algorithms from a more abstract point of view seems to originate with the work of Dilworth, Kutzarova and Temlyakov [4].

Most of this article will focus on greedy bases, but in the final section we will try to draw the reader's attention to some other intriguing questions that concern convergence of greedy algorithms.

2. Greedy and quasi-greedy bases

In [15], Konyagin and Temlyakov introduced the formal notions of greedy and quasi-greedy bases in a Banach space. Suppose X is a (real) Banach space and $(e_n)_{n=1}^{\infty}$ is a Schauder basis of X. We will denote by $(e_n^*)_{n=1}^{\infty}$ the corresponding bi-orthogonal functions. Assume that $(e_n)_{n=1}^{\infty}$ is normalized, i.e. $\|e_n\| = 1$ for all n. Each $x \in X$ has a unique expansion

$$x = \sum_{n=1}^{\infty} e_n^*(x)e_n.$$

We will take these terms in decreasing order of magnitude; however, when two terms are of equal size we take them in the basis order. Thus for each n we let A_n be the unique subset of $\mathbb{N}$ of n elements with the properties that:

$$j \in A_n,\ k \notin A_n \implies |e_k^*(x)| < |e_j^*(x)| \quad \text{or} \quad |e_k^*(x)| = |e_j^*(x)| \text{ and } j \leq k.$$

We define the *n-term greedy approximation* to x by

$$\mathcal{G}_n(x) = \sum_{j \in A_n} e_j^*(x)e_j.$$

Notice the operator $\mathcal{G}_n$ is highly nonlinear in x.

Now we define $(e_n)_{n=1}^{\infty}$ to be *quasi-greedy* if $\lim_{n\to\infty} \|\mathcal{G}_n(x) - x\| = 0$ for all $x \in X$. This means of course that the greedy approximations do converge, and is essentially the minimal requirement we might have to make this a reasonable method of approximation.

Of course, any unconditional basis is quasi-greedy. Let us give a simple example of a non-quasi-greedy basis. Let $(s_n)_{n=1}^{\infty}$ be the summing basis in c_0. Then $\sum_{n=1}^{\infty} \xi_n s_n$ converges if and only if the (real) series $\sum_{n=1}^{\infty} \xi_n$ converges. Let

$$\xi_n = \left(1, -1, \frac{1}{2}, -\frac{1}{4}, -\frac{1}{4}, \frac{1}{3}, -\frac{1}{9}, -\frac{1}{9}, -\frac{1}{9}, \dots\right)$$

so that each $1/n$ is followed by n terms $-1/n^2$. Then

$$\left\| \mathcal{G}_{3n^2-n/2}\left(\sum_{k=1}^{\infty} \xi_k s_k\right)\right\| \geq \sum_{k=1}^{n^2} \frac{1}{k} - \sum_{k=1}^{n} \frac{1}{k} \to \infty.$$

At the opposite extreme we may ask if it is true that the greedy algorithm in this instance gives the *best* method of approximating each $x \in X$. Konyagin and Temlyakov formalized this notion as follows. For each $x \in X$ and $n \in \mathbb{N}$ we define

$$\sigma_n(x) = \inf \left\{ \|x - \sum_{j\in A} \alpha_j e_j\| : \qquad |A| = n,\ \alpha_j \in \mathbb{R},\ j \in A \right\} \tag{2.1}$$

to measure the minimal possible error in an n-term approximation. Then we say that $(e_n)_{n=1}^{\infty}$ is *greedy* if there is a constant C so that

$$\|x - \mathcal{G}_n(x)\| \leq C\sigma_n(x) \qquad x \in X.$$

Thus, up to a constant C, one cannot improve on the greedy method of approximation.

If we have a basis which is not normalized, then we use the terms quasi-greedy and greedy if the corresponding normalized basis has these properties. Before proceeding, let us note that these definitions *seem* isometric in nature: they depend on the normalization of $(e_n)_{n=1}^{\infty}$ and so, passing to an equivalent norm (and then renormalizing $(e_n)_{n=1}^{\infty}$, should disrupt the behavior of the algorithm). This is not the case, however, and one can show that both definitions are invariant for equivalent norms.

We define a basis $(e_n)_{n=1}^{\infty}$ to be *democratic* if there is a constant Δ so that if A and B are two finite subsets of $\mathbb{N}$ with $|A| \leq |B|$ then

$$\|\sum_{j\in A} e_j\| \leq \Delta \|\sum_{j\in B} e_j\|.$$

Democracy means essentially that any n basis vectors have the same weight as any other n basis vectors.

The main theorem here is

Theorem 2.1. *(Konyagin-Temlyakov [15]) A basis $(e_n)_{n=1}^{\infty}$ of a Banach space X is greedy if and only if it is unconditional and democratic.*

This theorem is a very satisfactory characterization of a greedy bases. Of course symmetric bases are greedy, but, more importantly there are practical and useful examples. The Haar basis of $L_p[0,1]$ is greedy if $1 < p < \infty$ for example. See also [22].

In the case of quasi-greedy bases, Wojtaszczyk [22] showed that a basis is quasi-greedy if and only if the greedy operators $\mathcal{G}_n$ are uniformly bounded i.e. for some constant C

$$\|\mathcal{G}_n(x)\| \leq C\|x\| \qquad x \in X, \ n = 1, 2, \ldots.$$

Recall that these operators are nonlinear so this is not just the Uniform Boundedness Principle! He was also able to show that a Hilbert space has a quasi-greedy basis which is not greedy (note that a basis of a Hilbert space is greedy if and only if it is unconditional). We also note that Dilworth and Mitra [5] constructed a conditional quasi-greedy basis of ℓ_1. A recent very interesting example of a quasi-greedy basis in a space of functions of bounded variation is given in [1].

3. Almost greedy bases and duality

Let us look again at the definition of a greedy basis. It is important to notice that the definition of the optimal error $\sigma_n(x)$, (2.1), involves arbitrary coefficients. It would not be the same to consider

$$\tilde{\sigma}_n(x) = \inf \left\{ \|x - \sum_{j \in A} e_j^*(x) e_j\| : \qquad |A| = n, \ \alpha_j \in \mathbb{R}, \ j \in A \right\} \tag{3.2}$$

where one puts the expected coefficients $e_j^*(x)$ in place of the α_j. The author found this out quite by accident attempting to prove the Konyagin-Temlyakov theorem in a seminar with the wrong definition!

So we may define a basis $(e_n)_{n=1}^{\infty}$ to be *almost greedy* if there is a constant C so that

$$\|x - \mathcal{G}_n(x)\| \leq C\tilde{\sigma}_n(x) \qquad x \in X.$$

It turns out, rather remarkably, that this definition has some nice equivalent formulations [3]:

Theorem 3.1. *Let X be a Banach space with a basis $(e_n)_{n=1}^\infty$. The following conditions are equivalent:*
(i) $(e_n)_{n=1}^\infty$ is almost greedy.
(ii) $(e_n)_{n=1}^\infty$ is quasi-greedy and democratic.
(iii) For some (respectively, every) $\lambda >$ there is a constant $C = C_\lambda$ so that

$$\|x - \mathcal{G}_{[\lambda n]}(x)\| \leq C\sigma_n(x).$$

Here (iii) says that the greedy algorithm is essentially the best if one allows a small percentage increase in n, so the terminology is justified. There is in fact very little difference between the theory of almost greedy basis and that of greedy bases, except that an almost greedy basis need not be unconditional. However in a Hilbert space it is not difficult to show that any quasi-greedy basis is already almost greedy.

We now consider duality. If $(e_n)_{n=1}^\infty$ is a basis of X then $(e_n^*)_{n=1}^\infty$ is a basic sequence in X^* and if X is reflexive it is a basis for X^*. How do the three properties we have introduced dualize? In fact the Dilworth-Mitra example [5] of a quasi-greedy basis of ℓ_1 has the property that the dual basic sequence is not quasi-greedy [3]; in fact this example is almost greedy as well. Similarly, Oswald [19] showed that the Haar basis of H_1 is greedy while the dual basic sequence in BMO fails to be greedy; of course the dual sequence is unconditional and hence quasi-greedy in this instance.

Nevertheless there are positive results available [3]:

Theorem 3.2. *Let X be a Banach space with non-trivial Rademacher type. Suppose $(e_n)_{n=1}^\infty$ is an (almost) greedy basis of X; then $(e_n^*)_{n=1}^\infty$ is an almost greedy basic sequence in X^*.*

The counterexamples to such a theorem in general depend on the *fundamental function*

$$\varphi(n) = \|\sum_{j=1}^n e_j\|.$$

If the fundamental function increases slowly enough then the properties of being greedy or almost greedy pass to the dual. For a precise formulation we refer to [3].

4. Existence of quasi-greedy and almost greedy bases

It is clear that a Banach space cannot have a greedy basis unless it has an unconditional basis and this rules out many natural examples such as

$L_1[0,1]$ and $\mathcal{C}[0,1]$. One can also gives other examples of spaces failing to have a greedy basis such as $\ell_1 \oplus \ell_2$; here it is a classical result of Edelstein and Wojtaszczyk [7] that any normalized unconditional basis is equivalent to the canonical basis, which is plainly not democratic.

However it much easier to have an almost greedy basis and a very general construction was given in [2]. We start from the assumption that X has a basis, and then suppose that it has a complemented subspace S with a symmetric basis. Then X is isomorphic to $X \oplus S$. We now construct a new basis of the direct sum $X \oplus S$ which behaves very like the symmetric basis of S, in the sense that it inherits many of the good properties of the symmetric basis. The following theorem is proved in [2].

Theorem 4.1. *Let X be a Banach space with a basis and suppose X has a complemented subspace S with a symmetric basis. Then:*
(i) If S is not isomorphic to c_0 then X has a quasi-greedy basis.
(ii) If S has finite cotype then X has an almost greedy basis.

From this theorem one gets immediately that L_1 has an almost greedy basis, for example. It also appears that there is an obstruction around c_0. Of course c_0 has a greedy basis in the canonical basis, but we saw earlier that the summing basis is not quasi-greedy.

At this point we should observe the connection between quasi-greediness and unconditionality. A quasi-greedy basis need not be unconditional of course, but it preserves some vestige of unconditionality. For example there is a constant C so that we have

$$\|\sum_{j=1}^{n} \epsilon_j a_j e_j\| \leq C \|\sum_{j=1}^{n} a_j e_j\|$$

whenever $\epsilon_j = \pm 1$ *provided* we make the requirement that all the non-zero coefficients (a_j) are approximately of the same size, e.g. $1 \leq |a_j| \leq 2$ if $a_j \neq 0$. This is an important restriction and it turns out that classical arguments in Banach space theory can be used to show it is not always possible to find such a basis.

Recall the classical result of Lindenstrauss and Pełczyński [16]:

Theorem 4.2. *Any normalized unconditional basis of c_0 (or any $\mathcal{L}_\infty$-space) is equivalent to the canonical basis of c_0.*

Based on the same ideas, one can prove [2]:

Theorem 4.3. *Let X be a $\mathcal{L}_\infty$-space; if X has a quasi-greedy basis $(e_n)_{n=1}^\infty$*

then X *is isomorphic to* c_0 *and* $(e_n)_{n=1}^{\infty}$ *is equivalent to the canonical basis of* c_0.

In particular, the space $\mathcal{C}[0,1]$ has no quasi-greedy basis.

5. Basic sequences: an open question

From the point of view of Banach space theory, it becomes natural to ask whether every Banach space at least contains a quasi-greedy basic sequence. In view of the remarks above, there is some connection here with the famous unconditional basic sequence problem, which was settled in the celebrated paper of Gowers and Maurey [11]. But it should be a lot easier to contain a quasi-greedy basic sequence. A natural related question is whether every weakly null sequence has a subsequence which is quasi-greedy; of course if this is true, then every Banach space has a quasi-greedy basic sequence by Rosenthal's theorem [20].

This problem was first examined in [2] and the *easy* case was settled:

Theorem 5.1. *Let* X *be a Banach space which does not have* c_0 *as a spreading model (e.g. suppose* X *has nontrivial cotype). Then every normalized weakly null sequence has a quasi-greedy subsequence.*

However the problem in general seems rather hard, and curiously intersects with ideas that have already been considered in Banach space theory. In 1978, Elton had considered a very similar problem [8]. Elton's theorem is the following:

Theorem 5.2. *For each* $0 < \delta < 1$ *there is a constant* $C(\delta)$ *with the following property. Let* $(x_n)_{n=1}^{\infty}$ *be a weakly null sequence in a Banach space: then* $(x_n)_{n=1}^{\infty}$ *has a subsequence* $(y_n)_{n=1}^{\infty}$ *such that*

$$\|\sum_{j\in A} a_j y_j\| \le C\|\sum_{j=1}^{\infty} a_j y_j\|$$

for all finitely nonzero sequences $(a_j)_{j=1}^{\infty}$ *such that* $\max_j |a_j| \le 1$ *and all* $A \subset \{j : \ |a_j| \ge \delta\}$.

Thus Elton's theorem gives a restricted form of unconditionality for subsequences of $(x_n)_{n=1}^{\infty}$. It is worth noting that at the time of Elton's work examples were already known of weakly null sequences which contain no unconditional basic sequence [18]; we also remark that there is a recent result of Johnson, Maurey and Schechtman [13] giving such an example in $L_1[0,1]$.

As pointed out in [2] Elton's argument gave a function $C(\delta) \sim \log(1/\delta)$. In order to find a quasi-greedy subsequence one needs that $C(\delta)$ is uniformly bounded. In a recent preprint [6] the authors give an in-depth examination of this problem and prove many results on *partial unconditionality* but the main question remains open.

Of course we have now come a long way from practical applications!

6. More general greedy algorithms

In the last section we consider a rather more general situation. Suppose X is a Banach space and we are given a subset X called a *dictionary* D. It is assumed that $d \in D$ implies $-d \in D$ and the closed linear span of D is X. We would like to approximate every $x \in X$ by a linear combination of the elements of D.

Let us suppose X is a Hilbert space. There is a natural greedy way to do this to achieve our approximation. Let us assume that the dictionary D is a *compact* set. If $x = x_1 \in X$ pick $d_1 \in D$ and $a_1 \geq 0$ so that $\|x_1 - a_1 d_1\|$ is minimized. Now let $x_2 = x_1 - a_1 d_1$ and repeat the procedure. This will develop a series $a_1 d_1 + a_2 d_2 + \cdots +$ and the natural question is whether this series converges to x.

Let us call this the *pure greedy algorithm* or (PGA). Note that in the (PGA) we inductively construct $(x_n)_{n=1}^{\infty}$, $(a_n)_{n=1}^{\infty}$ and $(d_n)_{n=1}^{\infty}$ by the rules:

$$a_n = (x, d_n) = \max_{d \in D}(x, d),$$

$$x_{n+1} = x_n - a_n d_n.$$

In general the dictionary is not a compact set and we must consider a variant of this algorithm involving a *weakness* parameter c. Let us formally describe the weak dual greedy algorithm (WDGA) in a Hilbert space. Fix $0 < c < 1$. Let $x = x_1$ and then choose $a_n \geq 0$, $d_n \in D$ and $x_n \in X$ so that

$$a_n = (x, d_n) > c \sup_{d \in D}(x, d),$$

$$x_{n+1} = x_n - a_n d_n.$$

The question is whether this procedure converges i.e. does $\lim_{n \to \infty} \|x_n\| = 0$ (regardless of the choices made along the way)? Then of course $x = \sum_{n=1}^{\infty} a_n d_n$. This procedure was first considered in the statistics literature [12] and convergence was proved by Jones [14].

Theorem 6.1. *If X is a Hilbert space then the weak dual greedy algorithm converges.*

Now let us consider what a similar algorithm would look like in an arbitrary Banach space. We suppose that X has a Gateaux smooth norm so that for each $x \neq 0$ there is a unique linear functional φ_x so that $\|\varphi_x\| = 1$ and $\varphi_x(x) = \|x\|$. Fix $0 < c < 1$ as before. Let $x = x_1$ and then choose $a_n \geq 0$, $d_n \in D$ and $x_n \in X$ so that

$$\varphi_{x_n}(d_n) > c \sup_{d \in D} \varphi_{x_n}(x),$$

$$\|x_n - a_n d_n\| = \min_{a \in \mathbb{R}} \|x_n - a d_n\|,$$

and

$$x_{n+1} = x_n - a_n d_n.$$

We call this the (WDGA) as before and the question is whether the (WDGA) converges, i.e. does $\lim_{n \to \infty} \|x_n\| = 0$.

This is an isometric question and should really depend on the choice of norm. It seems reasonable to restrict attention to spaces X which are uniformly convex or uniformly smooth or both. But even with these restrictions very little is known. However, Ganichev and the author [10] were able to extend the convergence from Hilbert spaces to the L_p-spaces where $1 < p < \infty$, although the result depends critically on rather delicate properties of the norm.

Theorem 6.2. *Let X be a subspace of a quotient of $L_p[0,1]$ where $1 < p < \infty$. Then the weak dual greedy algorithm converges in X.*

In a paper under preparation we will give some extensions of this result (including to the Schatten classes S_p where $1 < p < \infty$). However the general problem remains completely open. For example we do not know if the (WDGA) converges in every uniformly convex uniformly smooth space.

Of course the (PGA) makes sense in an arbitrary Banach space if the dictionary is compact. Let $x = x_1$ and then choose $a_n \geq 0$, $d_n \in D$ and $x_n \in X$ so that

$$\|x_n - a_n d_n\| = \min_{\substack{a \geq 0 \\ d \in D}} \|x_n - a d\|,$$

$$x_{n+1} = x_n - a_n d_n.$$

Unfortunately letting $c \to 0$ in the (WDGA) does not reduce to this algorithm then.

Little seems to be known about the (PGA). Livshits [17] gives an example of a space with a Gateaux smooth norm where this algorithm fails to converge.

A natural weak form of the (PGA) is as follows. Fix $0 < c < 1$ as before. Let $x = x_1$ and then choose $a_n \geq 0$, $d_n \in D$ and $x_n \in X$ so that

$$\|x_n - a_n d_n\|^2 \leq (1 - c^2)\|x_n\|^2 + c^2 \inf_{\substack{a \geq 0 \\ d \in D}} \|x_n - ad\|^2,$$

and

$$x_{n+1} = x_n - a_n d_n.$$

Let us call this the weak greedy algorithm (WGA). When does this algorithm converge?

A different variant of the (PGA) has been considered by Livshits [17] and he proves convergence under essentially the same hypotheses as Theorem 6.2 (taking into account results in [10]).

This whole subject appears wide open for exploration. These algorithms and many more are considered in the survey [21], which is a good place for the interested reader to start.

References

[1] P. Blecher, R. DeVore, A. Kamont, G. Petrova, and P. Wojtaszczyk, *Greedy wavelet projections are bounded on BV*, Trans. Amer. Math. Soc., to appear.

[2] S. J. Dilworth, N. J. Kalton, and D. Kutzarova, *On the existence of almost greedy bases in Banach spaces*, Studia Math. **159** (2003), no. 1, 67–101. Dedicated to Professor Aleksander Pełczyński on the occasion of his 70th birthday.

[3] S. J. Dilworth, N. J. Kalton, D. Kutzarova, and V. N. Temlyakov, *The thresholding greedy algorithm, greedy bases, and duality*, Constr. Approx. **19** (2003), no. 4, 575–597.

[4] S. J. Dilworth, D. Kutzarova, and V. N. Temlyakov, *Convergence of some greedy algorithms in Banach spaces*, J. Fourier Anal. Appl. **8** (2002), no. 5, 489–505.

[5] S. J. Dilworth and D. Mitra, *A conditional quasi-greedy basis of l_1*, Studia Math. **144** (2001), no. 1, 95–100.

[6] S. J. Dilworth, E. Odell, T. Schlumprecht, and A. Zsak, *Partial unconditionality*, preprint.

[7] I. S. Èdel′šteĭn and P. Wojtaszczyk, *On projections and unconditional bases in direct sums of Banach spaces*, Studia Math. **56** (1976), 263–276.

[8] J. Elton, *Weakly null normalized sequences in Banach spaces*, Ph.D. thesis, Yale Univ., 1978.

[9] M. Frazier, B. Jawerth, and G. Weiss, *Littlewood-Paley theory and the study of function spaces*, CBMS Regional Conference Series in Mathematics, vol. 79, Published for the Conference Board of the Mathematical Sciences, Washington, DC, 1991.

[10] M. Ganichev and N. J. Kalton, *Convergence of the weak dual greedy algorithm in L_p-spaces*, J. Approx. Theory **124** (2003), no. 1, 89–95.

[11] W. T. Gowers and B. Maurey, *The unconditional basic sequence problem*, J. Amer. Math. Soc. **6** (1993), 851–874.

[12] P. J. Huber, *Projection pursuit*, Ann. Statist. **13** (1985), no. 2, 435–525.

[13] W. B. Johnson, B. Maurey, and G. Schechtman, *Weakly null sequences in L_1*, preprint.

[14] L. K. Jones, *On a conjecture of Huber concerning the convergence of projection pursuit regression*, Ann. Statist. **15** (1987), no. 2, 880–882.

[15] S. V. Konyagin and V. N. Temlyakov, *A remark on greedy approximation in Banach spaces*, East J. Approx. **5** (1999), 365–379.

[16] J. Lindenstrauss and A. Pełczyński, *Absolutely summing operators in L_p-spaces and their applications*, Studia Math. **29** (1968), 275–326.

[17] E. D. Livshits, *Convergence of greedy algorithms in Banach spaces*, Mat. Zametki **73** (2003), no. 3, 371–389 (Russian, with Russian summary).

[18] B. Maurey and H. P. Rosenthal, *Normalized weakly null sequence with no unconditional subsequence*, Studia Math. **61** (1977), no. 1, 77–98.

[19] P. Oswald, *Greedy algorithms and best m-term approximation with respect to biorthogonal systems*, J. Fourier Anal. Appl. **7** (2001), no. 4, 325–341.

[20] H. P. Rosenthal, *A characterization of Banach spaces containing l^1*, Proc. Nat. Acad. Sci. U.S.A. **71** (1974), 2411–2413.

[21] V. N. Temlyakov, *Nonlinear methods of approximation*, Found. Comput. Math. **3** (2003), 33–107.

[22] P. Wojtaszczyk, *Greedy algorithm for general biorthogonal systems*, J. Approx. Theory **107** (2000), 293–314.

ON THE HAHN-BANACH THEOREM

LAWRENCE NARICI
Mathematics Department
St. John's University
Jamaica, NY 11439, USA
narici@gmail.com

Dedicated to my friend Jean Schmets
on the occasion of his retirement

I love the Hahn-Banach theorem. I love it the way I love *Casablanca* and the Fontana di Trevi. It is something not so much to be read as fondled. What is "the Hahn-Banach theorem?" Let f be a continuous linear functional defined on a subspace M of a normed space X. Take as the Hahn-Banach theorem the property that f can be extended to a continuous linear functional on X without changing its norm. Innocent enough, but the ramifications of the theorem pervade functional analysis and other disciplines (even thermodynamics!) as well. Where did it come from? Were Hahn and Banach the discoverers? The axiom of choice implies it, but what about the converse? Is Hahn-Banach equivalent to the axiom of choice? (No.) Are Hahn-Banach extensions ever unique? They are in more cases than you might think, when the unit ball of the dual is "round," as for ℓ_p with $1 < p < \infty$, for example, but not for ℓ_1 or ℓ_∞. Instead of a linear functional, suppose we substitute a normed space Y for the scalar field and consider a continuous linear map $A : M \to Y$. Can A be continuously extended to X with the same norm? Well, sometimes. Unsurprisingly, it depends on Y, more specifically, on the "geometry" of Y: If the unit ball of Y is a "cube," as for $Y = (\mathbf{R}^n, \|\cdot\|_\infty)$ or $Y =$ real ℓ_∞, for example, then for any subspace M of any X, any bounded linear map $A : M \to Y$ can be extended to X with the same norm. This is not true if $Y = \left(\mathbf{R}^n, \|\cdot\|_p\right)$, $n > 1$, for $1 < p < \infty$, despite the topologies being identical. The cubic nature of the unit ball does not suffice, however—if $Y = c_0$, the extendibility dies. This article traces the evolution of the analytic form as well as subsequent developments up to 2004.

1. What is it?

The two principal versions of the Hahn-Banach theorem are as a continuous extension theorem (analytic form) and as a separation theorem (geometric form) about separating convex sets by means of a continuous linear functional that takes different values on the sets.

ANALYTIC FORMS

Dominated version. Let f be a continuous linear functional defined on a subspace M of a real vector space (no norm) X, p a sublinear functional defined on X and $f \leq p$ on M; f can be extended to a linear functional F defined on X with $F \leq p$.

$$\begin{array}{lll} F : X & & F \leq p \\ \quad\ \ | & \searrow & \\ f : M & \longrightarrow \mathbf{R} & f \leq p \end{array}$$

For complex spaces, we mainly need some absolute values: If X is complex, and p a *seminorm* such that $|f| \leq p$ on M then $|F| \leq p$.

Norm-preserving version. If X is normed space over $K = \mathbf{R}$ or $\mathbf{C}$ and $f : M \longrightarrow K$ is a continuous linear functional then there exists a continuous linear functional F extending f defined on all of X such that $\|F\| = \|f\|$.

GEOMETRIC FORM Let X be a real or complex topological vector space. In any real or complex TVS X, if the linear variety $x + M$ does not meet the open convex set G then there exists a closed hyperplane H containing $x+M$ that does not meet G either. Mazur 1933 deduced the geometric form from the analytic form; he made no mention of the converse possibility. In a 1941 article, Dieudonné [1981b] refers to the geometric form as *the* Hahn-Banach theorem, so he was apparently aware of the equivalence of the two. It is first called the geometric form by Bourbaki.

The analytic form is a cousin of Tietze's theorem that a bounded continuous $f : K \to [a, b]$ defined on a closed subset K of a normal space T possesses a continuous extension $F : T \to [a, b]$ with the same bounds. The geometric form resembles Urysohn's lemma about separating disjoint closed subsets of a normal space by a continuous function. Urysohn's lemma is usually proved by induction, the geometric Hahn-Banach theorem by *transfinite* induction.

In the interest of keeping size reasonable, I consider only the analytic form in the sequel. There are many—denumerably, I suspect—other versions of the theorem, for vector lattices, modules, boolean algebras, bilinear functionals, groups, semigroups and more. It has many applications not only outside functional analysis but outside mathematics. Feinberg and Lavine [1983], for example, develop thermodynamics using the Hahn-Banach theorem, Neumann and Velasco [1994] apply Hahn-Banach type theorems to develop feasibility results on the existence of flows and potentials and Delbaen and Schachermayer [1994] use it to develop a fundamental theorem of asset pricing.

2. The Obvious Solution

Suppose that X is just a vector space—no norm—over $K = \mathbf{R}$ or $\mathbf{C}$ and the linear functional f maps a subspace M into K. An easy way to extend f to X is to take an algebraic complement N of M, consider the projection P_M on M along N and take $F = f \circ P_M$. In effect, take F to be 0 outside M. Will this technique work for *continuous* linear functionals f defined on a closed (extend f by continuity to $\overline{M}$ if M is not closed) subspace M of a topological vector space X? If P_M is continuous, then $F = f \circ P_M$ is a continuous linear extension of f though not necessarily of the same norm (cf. Sec. 7.1). Generally, however, we cannot rely on this method because P_M is continuous if and only if N is a *topological* complement of M [Narici and Beckenstein 1985, (5.8.1)(a)] and uncomplemented subspaces are common—c_0, for example, is an uncomplemented subspace of ℓ_∞, so there is no continuous projection of ℓ_∞ onto c_0 [Narici and Beckenstein 1985, Ex. 5.8.1]. $C[0,1]$, $L_p[0,1]$ and ℓ_p, $1 \leq p \leq \infty$, $p \neq 2$, have closed uncomplemented subspaces [Köthe 1969, pp. 430-1] and for $0 < p < 1$, no finite-dimensional subspace of $L_p[0,1]$, has a topological complement [Köthe 1969, p. 158]. In fact, any Banach space X has uncomplemented closed subspaces unless X is linearly homeomorphic to a Hilbert space [Lindenstrauss and Tzafriri 1971]. Some instances in which a subspace M of a locally convex space is complemented are M finite-dimensional or codimensional, or M a closed subspace of a Hilbert space, in which case its orthogonal complement $M^\perp$ is a topological complement. We say a little more about the Hilbert space situation in Sec. 7.2, this being a case in which $f \circ P_M$ is the *only* continuous linear extension of f with the same norm.

3. The Times

Throughout the nineteenth and early twentieth centuries the function concept was significantly broadened, analysis became "geometrized," the idea of structure emerged and the standards of rigor improved greatly; the techniques of Euclidean geometry became the standard. Functional analysis evolved from the desire to do analysis on function spaces, treating functions the way points in $\mathbf{R}$ or $\mathbf{R}^2$ had been dealt with.

3.1. *The Evolution of the Function Concept; Analysis on Spaces of Functions*

Fourier series played an important role not only in the enlargement of the function concept, but in analysis in general throughout the 19th century. Dirichlet boldly proffered the characteristic function of the rationals, the *Dirichlet function,* in 1829 as a function for which the integrals for the coefficients of its Fourier series "lose every meaning." There was no disagreement about the lack of integrability, but was this a function? something which could not be graphed? At the time, with no formal definition of function—indeed, with practically no formal definitions of anything—"function" had only its intuitive meaning. It was tacitly required that it had to be graphable and it meant essentially "elementary function:" polynomials, trigonometric functions, exponentials and logarithms. In 1837 Dirichlet proposed that *any* correspondence between the points of interval $[a, b]$ and points of $\mathbf{R}$ be considered a function. In view of the strange behavior of even continuous functions such as Riemann's continuous-for-irrational x, discontinuous-for-rational-x in 1854 and Weierstrass's nowhere differentiable continuous function in 1874, it became apparent that more latitude was clearly necessary. The ability to uniformly approximate "strange" functions by trigonometric functions helped render them acceptable as did Weierstrass's 1885 demonstration that any continuous function could be uniformly approximated by polynomials.

Consideration of functions whose domains were other than subsets of $\mathbf{R}$ or $\mathbf{C}$ has a venerable history. Circa 100 B.C. Zenodorus considered the isoperimetric problem—among the closed plane curves of a given length, find the one that encloses the most area—so, even at this early date there was consideration of numbers associated with curves and choosing the curve or curves corresponding to the least number. A similar situation occurred when the Bernoulli brothers considered the brachistochrone problem—from the class of curves connecting two points, associate a number, a time, with each and choose a curve corresponding to the least number. Throughout the 1700's it was common to associate numbers with curves by means of definite integrals. Not only was mapping curves into numbers common, since the early 1800's so were function-to-function mappings such as differential operators, Laplace transforms, and shift operators. In spite of this long history of linking functions with numbers or other functions, it took until the late 1800's to formalize the notion of a function as a correspondence between elements of arbitrary sets. ["Set" or "class" of functions had been in common use in the early 19th century, well before Boole in 1847 or Cantor. Volterra 1887 spoke of numerical- and $\mathbf{R}^n$-valued functions defined on the set of all continuous curves (*linee* or *lignes*) in a square and then did something decisively different: he proposed doing analysis on them—limits, continuity, derivatives. This was possible because there were already several distinct ways of judging proximity of functions that arose from the various notions of convergence of a sequence of functions that developed in the latter half of the 19th century. Volterra called these new kinds of functions *funzioni dipenditi da linee* or *fonctions de ligne* where by *ligne* he meant a continuous image of $[0, 1]$ in the unit square. Peano realized early the exotic possibilities of such a broad criterion for "curve." In part to make his point, he invented his space-filling curve. Undaunted by this cautionary example, Hadamard pursued it. In 1902 he wrote a short note [Bull. Soc. Math. France **30**, 40–43] on Volterra's derivatives of *fonctions de lignes.* In 1903 [Comptes Rendus **136**, 351–354] He abandoned *fonction de ligne* and called the new functions of functions *fonctionnelles*, analysis on them *analyse fonctionnelle* and gave our subject its name. Hadamard's student Paul Lévy wrote a book, *Leçons d'analyse fonctionnelle,* in 1922 in which he divided *calcul fonctionnelle* into *algèbre fonctionnelle* and *analyse fonctionnelle*; the *algèbre* dealt with problems whose unknowns were ordinary functions, the *analyse* with problems in which the unknowns were *fonctionnelles.*

3.2. *Structure and Isomorphism*

Throughout the 19th century, the idea that concretely different things could be the same in some crucial sense gestated, i.e., the notion of isomorphism. In what appears to be the first use of the term, the German chemist Eilhard Mitscherlich formulated the *principle of isomorphism* of crystals in 1819, concerning similarity of geometric forms of crystal structures and the chemical consequences of such an arrangement. Unlike our usage of *isomorphic,* chemists say *isomorphous* as in sodium nitrate and calcium sulfate have *isomorphous* crystal structure.

Contemporaneously, geometers such as Gauss, Lobachevsky and Bolyai—and Klein's Erlanger program—created non-Euclidean geometries and reformulated classical geometry. These developments influenced the idea of "space." Hilbert's 1899 *Grundlagen der Geometrie,* and its many subsequent editions, launched the abstract mathematics of the 20th century. The qualitative leap that Hilbert and others made is that they did not try to define points and lines and planes as Euclid had attempted; rather, they accepted these notions as "atoms," without intrinsic content. Not only didn't you know what they were, you couldn't. In a 1941 letter to Frege, Hilbert wrote:

> If among my points I consider some systems of things (e.g., the system of love, law, chimney sweeps . . . .) and then accept only my complete axioms as the relationships between these things, my theorems (e.g., the Pythagorean) are valid for these things also.

Through the medium of the axiomatic system, mathematics was about to attain a level of abstraction hitherto unknown.

The idea of a *vector* had been around in the 19th century but it meant n-tuple. With infinite-dimensional vector spaces in mind, in Chapter IX of his 1888 book, Peano gave a rather modern axiomatic definition of *vector space* and *linear map.* Pincherle had already been writing about *spazio funzionale, operazioni funzionali, calcolo funzionale* and linear operators on complex sequence spaces, so he was quite receptive to Peano's ideas. Pincherle wrote a book about vector spaces (*Le operazioni distributive e le loro applicazioni all'analisi* with U. Amaldi) in 1901, but the idea was mostly ignored. Even Hilbert in his research on ℓ_2 was indifferent to its vector space structure. It was not until Riesz, Helly, Hahn, Banach and Wiener endowed vector spaces with a norm that interest in them ignited. More generally, the idea of *structure* had arrived. *Group* (a term coined by Galois) was defined in 1895, *field* in 1903.

A comprehensive framework for the various notions of limit and continuity for particular sets of functions that Volterra and Hadamard were investigating materialized with M. Fréchet's abstract metric space in 1904, ideas refined in his 1906 thesis. Fréchet used Jordan's term, *écart,* for metric distance. Although Fréchet liberally used spatial imagery, he did not coin the term metric *space*; Hausdorff (probably) was the first to do so in his *Mengenlehre* [1914].

3.3. *Spatial Imagery and the Euclidean Renaissance*

By the late nineteenth and early twentieth centuries, suggestive geometric terminology and spatial imagery were commonly applied to arbitrary sets of 'points.' Hilbert and his school spoke of *orthogonal expansions;* in 1913, Riesz described the solution of systems of homogeneous equations

$$f_i(x) = a_{i1}x_1 + \cdots + a_{in}x_n = 0, \quad 1 \leq i \leq n$$

as an attempt to find $x = (x_1, \ldots, x_n)$ *orthogonal to* the linear span $[f_1, ..., f_n]$ where $f_i = (a_{i1}, ..., a_{in})$, i.e., solving the equations is an attempt to identify the orthogonal complement of $[f_1, ..., f_n]$. Significantly, the 'equations,' the f_i, achieved vector status; they were of the same species as the 'variables.' Peano's *Calcolo Geometrico* in 1888 and Minkowski's *Geometrie der Zahlen* in 1896 are two seminal works in the geometrizing of analysis. Minkowski tinctured analysis with ideas about *convexity* in $\mathbf{R}^n$. (Even so, in a 1941 article on the Hahn-Banach theorem mentioned earlier, Dieudonné 1981b *defines* convexity, so it wasn't standard even by 1941.) Minkowski defined support hyperplane, support function and proved the existence of a support hyperplane at every boundary point of a convex body. Helly extended Minkowski's notions about convexity from $\mathbf{R}^n$ to normed sequence spaces (see Sec. 4.4).

Contemporaneously, Euclidean methodology became established: theorem–proof arguments to make deductions from explicitly stated assumptions, the type of rigor that became prevalent in the 20th and early 21st centuries. The standards were raised so much that most earlier work looks shabby by comparison. For example, from the 17th into the 19th century, infinite series were usually treated the same as polynomials with no regard for convergence.

4. Origins

Attempts to solve infinite systems of linear equations led to early versions of the Hahn-Banach theorem as well as to the creation of the general normed space. The analog of the diagonalizability conditions for finite systems of linear equations, the necessary and sufficient condition for solvability of an infinite system of linear equations, is compatibility between the linear equations and the scalars which can be described as the continuity of a certain linear functional. (As normed spaces had not been defined yet, this was not the interpretation given at the time.) A key figure is F. Riesz (no surprise) who proved variants of the Hahn-Banach theorem for $L_p[0,1]$, ℓ_p and $BV[a,b]$. In the course of his investigations with $C[a,b]$, Riesz very nearly defined the general normed space. Helly also proved special cases of the Hahn-Banach theorem, defined a general normed *sequence* space and a dual of a sequence space. Hahn, Banach and Wiener subsequently defined the general real normed space. Hahn and Banach each independently proved the Hahn-Banach theorem for real normed spaces..

4.1. *Systems of Linear Equations*

Two prominent problems of the late nineteenth century were:
Moment problems. Given a sequence (c_n) of numbers, find a function x with those 'moments,' i.e., such that

$$f_n(x) = \int_0^1 t^n x(t)\,dt = c_n \text{ for every } n \in \mathbf{N}$$

Fourier series. Given a sequence (g_n) of cosines or sines and (c_n) of numbers, find a function x for which the c_n are the Fourier coefficients, i.e., x such that

$$f_n(x) = \int_{-\pi}^{\pi} x(t)g_n(t)\ \ dt = c_n \text{ for every } n \in \mathbf{N}$$

We can rephrase these in a more general setting. Let X be a normed space with dual X', let S be a set, and let $\{c_s : s \in S\}$ be a collection of scalars.

(V) The vector problem. Let $\{f_s : s \in S\}$ be a collection of bounded linear functionals on X. Find $x \in X$ such that $f_s(x) = c_s$ for every s.
and its dual:

(F) The functional problem. Let $\{x_s : s \in S\}$ be a collection of vectors from X. Find $f \in X'$ such that $f(x_s) = c_s$ for every s.

If X is reflexive then solving (F) also solves (V), for given "vectors" $\{f_s : s \in S\} \subset X'$ there exists $h \in X''$ such that $h(f_s) = c_s$ for every s. Now choose $x \in X$ such that $h(f_s) = f_s(x)$ for every s.

Suppose X is a *reflexive* normed space and consider a simple vector problem: Given functionals $f, g \in X'$ and scalars a and b, find $x \in X$ such that $f(x) = a$ and $g(x) = b$. If f and g are linearly independent, for any scalars c and d, take $h(cf + dg) = ca + db$. Then extend the continuous linear functional h to $H \in X''$ by the Hahn-Banach theorem. Finally, choose $x \in X$ such that for all $\phi \in X'$, $H(\phi) = \phi(x)$ to solve the problem.

If f and g are linearly *dependent,* then the scalars have to be "compatible." If $g = 2f$, say, then we must have $b = 2a$. More generally, given functionals $f_1, f_2, \ldots, f_n$ and scalars $c_1, c_2, \ldots, c_n$, if $\sum a_i f_i = 0$ then we must also have $\sum a_i c_i = 0$. This type of compatibility is guaranteed by conditions (*) and (**) in Sec. 4.2 and (***) of Theorem 4.1 where it can be viewed as a continuity condition.

4.2. *Riesz*

Motivated by Hilbert's work on $L_2[0,1]$, Riesz 1910 invented the spaces $L_p[0,1]$, $1 < p < \infty$; in 1913 he considered the ℓ_p spaces. He generalized the moment and Fourier series problems [1910, 1911] to the vector problem (**LP**) below. In solving (**LP**), he inadvertently solved a functional problem and created an early Hahn-Banach extension theorem. Riesz solved the vector problem in the reflexive spaces $L_p[0,1]$, $p > 1$. Simultaneously, he solved an associated functional problem in $L_p' = L_q$ which yielded a special case of the Hahn-Banach theorem.

(**LP**) Let S be a set. For $p > 1$ and $1/p + 1/q = 1$, given y_s in $L_q[a,b]$ [equivalently, consider the functionals f_s in Eq. (1)] and scalars c_s for each $s \in S$, find x in $L_p[a,b]$ such that

$$f_s(x) = \int_a^b x(t) y_s(t) dt = c_s \text{ for every } s \tag{1}$$

For there to be such an x, he showed that the following necessary and sufficient connection between the y's and the c's had to prevail: There exists $K > 0$ such that for any finite set of indices s and scalars a_s

$$\left|\sum a_s c_s\right| \leq K \left(\int_a^b \left|\sum a_s y_s\right|^q\right)^{1/q} = K \left\|\sum a_s y_s\right\|_q \tag{*}$$

in today's language. Condition (*) means that if the y's are linearly dependent, if $\sum a_s y_s = 0$ for some finite set of scalars a_s, then $\sum a_s c_s = 0$ as well. Thus, if we consider the linear functional g on the linear span $M = [y_s : s \in S]$ of the y's in $L_q[a,b]$ defined by taking $g(y_s) = c_s$, the g so obtained is well-defined. Not only that, for any y in M, $|g(y)| \leq K \|y\|_q$ on M, so g is continuous on M. If there is an x in L_p which solves (**LP**), then g has a continuous extension G to L_q, namely, for any y in L_q,

$$G(y) = \int x(t) y(t)\; dt \quad [G(y_s) = c_s \quad (s \in S)]$$

Thus, Riesz showed that (**LP**) is solvable if and only if a certain linear functional g defined on a subspace of L_q is continuous; if it is solvable, g is also special in that it is a restriction of a continuous linear functional defined on all of L_q.

In a paper written and submitted in 1916 but not published until 1918, Riesz turned to the following vector problem.

(**BV**) Given $y_s \in C[a,b]$, and scalars c_s $(s \in S)$, find $x \in BV[a,b]$ (functions of bounded variation on $[a,b]$) such that

$$f_s(x) = \int_a^b y_s(t)\ dx(t) = c_s \quad (s \in S)$$

He solved it with a necessary and sufficient condition—continuity, again—very much like (*), namely: There exists $K > 0$ such that for any finite set of indices s and any scalars a_s

$$\left|\sum a_s c_s\right| \leq K \sup{}_{t\in[a,b]} \left|\sum a_s y_s\,(t)\right| = K \left\|\sum a_s y_s\right\|_\infty \qquad (**)$$

in modern notation.

4.3. *The First Normed Space*

In his 1918 article, Riesz used "norm" and the notation $\|x\| = \sup|x|\,[0,1]$ for x in the *funktionraum* $C\,[0,1]$. He observed that

> $\|x\|$ is generally positive, and is zero only when $x\,(t)$ vanishes identically. Furthermore ...(for any scalar a and any $x, y \in C\,[0,1]$)
>
> $$\|ax\,(t)\| = |a|\,\|x\|\,,\ [\text{and}]\ \|x+y\| \leq \|x\| + \|y\|$$
>
> By the distance of x, y we understand the norm $\|x-y\| = \|y-x\|$.

He didn't define the general normed space but he came mighty close. Although he was working in particular spaces, Riesz intuited that these were only galaxies in a greater universe. Even in 1913, in the introduction to his book, *Les systèmes d'équations linéaires à une infinité d'inconnues,* he said:

> Strictly speaking, our study is not part of the theory of functions but rather can be considered as a first stage of a theory of functions of infinitely many variables.

The notation $\|\cdot\|$ was first used by Hilbert's student E. Schmidt 1908, as in "Gram-Schmidt," who called $x = (a_n) \in \ell_2$ "normiert" if $\sum |a_n|^2 < \infty$, then took $\|x\| = \left(\sum |a_n|^2\right)^{1/2}$. When he defined norm in his book, Banach used $|\cdot|$, not $\|\cdot\|$.

Riesz unintentionally extended some continuous linear functionals in solving (**LP**) and (**BV**). Helly made a qualitative leap: he went directly to the extension.

4.4. *Enter Helly*

Despite the brevity of his mathematical career—only five journal publications—the Austrian mathematician Eduard Helly (1884–1943) made significant contributions to functional analysis. Dieudonné 1981a [p. 130] characterized Helly's 1921 article as "a landmark in the history of functional analysis." For an excellent discussion of Helly and his work, see Butzer *et al.* 1980 and 1984. By means different from and simpler than Riesz's, Helly also solved (**BV**) in 1912 and proved special cases of the Hahn-Banach and Banach-Steinhaus theorems in $C[a,b]$. A bullet through the lungs in September 1915—a wound that ultimately caused his death—ended his stint at the eastern front as a soldier in the Austrian army in World War I. He spent almost the next five years as a prisoner of war, enduring eastern Siberia's frigidity since 1916. He did not return to Vienna until mid-November of 1920 by way of Japan, the Middle East and Egypt. I would like to think that the opportunity for distraction afforded by thinking about mathematics helped sustain him during that awful period. In any case he revisited (**BV**) in 1921 with a perspective that definitively anticipates the Hahn-Banach theorem. Probably in the belief that all spaces were reflexive, Helly tried to solve (**BV**) by means of a corresponding functional problem, then finding the "vector" (in X'', actually) that corresponded to the functional. He defined a general normed sequence space and a dual. Some highlights of his approach are:

►**Normed sequence spaces** Helly dealt with a general vector subspace X of the space $\mathbf{C}^{\mathbf{N}}$ of complex sequences equipped with a norm, an *abstandfunktion* or "distance function." He did not use the notation $\|\cdot\|$. This is general enough to cover the ℓ_p spaces and many others such as L_2 which can be identified with ℓ_2. Helly realized that a norm generalized what Minkowski (in $\mathbf{R}^n$) called the *gauge* of a convex body.

►**Dual spaces** Given a normed subspace X of $\mathbf{C}^{\mathbf{N}}$, Helly took as its "dual space"

$$X' = \left\{ (u_n) \in \mathbf{C}^{\mathbf{N}} \colon \sum_{n \in \mathbf{N}} x_n u_n < \infty \text{ for all } (x_n) \in X \right\}$$

i.e., (u_n) such that $(x_n u_n)$ is summable for all $(x_n) \in X$. If $X = c$ or c_0, then $X' = \ell_1$ by this method; if $X = \ell_1$, then $X' = \ell_\infty$. If $X = \ell_\infty$, the X' you get this way is only part of what we call the dual of X today.

►**Norms the dual space** For $x = (x_n) \in X$ and $u = (u_n) \in X'$, Helly defines a bilinear form $\langle \cdot, \cdot \rangle$ on $X \times X'$ (that makes (X, X') a dual pair) by taking for x in X and u in X' $\langle x, u \rangle = \sum_{n \in \mathbf{N}} x_n u_n$. Using an idea of Minkowski's, he considers a *dual norm* for X' by taking

$$\|u\| = \sup \left\{ \frac{|\langle x, u \rangle|}{\|x\|} : x \neq 0 \right\}$$

The dual norm on X obtained from X' by this technique yields the original norm on X. (Nowadays such pairs (X, X'), subject to *absolute* convergence of $\sum x_n u_n$, are called Köthe sequence spaces and α-duals, respectively.) The dual norm also permitted Helly to consider successive duals. He sought to solve the following vector problem:

(**SQ**) Given sequences $f_n = (f_{nj})$ from $X' \subset \mathbf{C}^{\mathbf{N}}$, and a sequence $(c_n) \in \mathbf{C}^{\mathbf{N}}$, find $x \in X$ such that

$$\langle x, f_n \rangle = \sum_{j \in \mathbf{N}} x_j f_{nj} = c_n \text{ for each } n \in \mathbf{N}$$

He tried to solve (**SQ**) by finding $h \in X''$ such that $h(f_n) = c_n$ and using reflexivity. He discovered that the $x \in X$ corresponding to h did not always exist, thus showing that some spaces are not reflexive. As part of his solution, with a condition like (*) above, Helly proved a restricted version of Theorem 4.1, below, by extending a bounded linear functional f from a subspace M to the whole space. The key step was, for x not in M, to find a linear F such that

$$\begin{array}{lccl} F: M \oplus \mathbf{C}x & & & F \leq k\,\|\cdot\| \\ & | & \searrow & \\ f: & M & \longrightarrow & K,\ f \leq k\,\|\cdot\| \end{array}$$

the same idea that Hahn 1927 and Banach 1929 used to prove the Hahn-Banach theorem.

4.5. *Hahn and Banach*

Riesz 1918 had considered $C[a,b]$ as a normed space. Hahn 1922, Wiener 1922 and Banach 1923 took the final step: each independently defined the general *real* normed space. Considering Hahn and Banach's awareness of what Helly did (complex normed sequence spaces), it is surprising that neither mentions complex scalars—indeed, Banach does not consider them even in his 1932 book. Wiener 1923 observed that complex scalars could be used just as well as real scalars. Hahn called a norm a *Massbestimmung.* Hahn and Banach each required completeness. Banach removed it in his book, distinguishing between normed and Banach spaces. Each considered the general problem of extending a continuous linear functional defined on a subspace of a general normed space, not a sequence space, as Helly did. Hahn published the norm-preserving form of the theorem in 1927; Banach proved it independently and published his result in 1929. He mentioned Helly, acknowledged Hahn's priority in his book 1932, and generalized it to the dominated version. Although he made no further use of the greater generality, the more general result was useful with the advent of locally convex spaces. Each of them used Helly's technique to obtain the theorem—reduce the problem to the case of a one-vector extension —but instead of ordinary induction, they used *transfinite* induction. (The Zorn's lemma equivalent of transfinite induction commonly used today did not arrive until 1935.) Hahn also formally introduced the notion of dual space (*polare Raum*), noted that X is embedded in its second dual X'' and defined reflexivity (*regularität*).

Generalizing Riesz's results for $L_p[0,1]$ and $BV[a,b]$, Banach 1932 extended a result of Helly 1912 to obtain the compatibility condition (***) of Theorem 4.1 below—usually referred to as *Helly's theorem*—and solved the general functional problem.

Theorem 4.1. (Helly-Banach) *Let X be a real (or complex) normed space, let $\{x_s\}$ and $\{c_s\}$, $s \in S$, be sets of vectors and scalars, respectively. Then there is a continuous linear functional f on X such that $f(x_s) = c_s$ for each $s \in S$ if and only if there exists $K > 0$ such that for all finite subsets $\{s_1, \ldots, s_n\}$ of S and scalars $a_1, \ldots, a_n$*

$$\left|\sum_{i=1}^{n} a_i c_{s_i}\right| \leq K \left\|\sum_{i=1}^{n} a_i x_{s_i}\right\| \qquad (***)$$

Banach used the Hahn-Banach theorem to prove Theorem 4.1 but Theorem 4.1 implies the Hahn-Banach theorem: Assuming that Theorem 4.1

holds, let $\{x_s\}$ be the vectors of a subspace M, let f be a continuous linear functional on M; for each $s \in S$, let $c_s = f(x_s)$. Since f is continuous, condition (***) is satisfied and f possesses a continuous extension to X.

5. The Complex Case

F. Murray 1936 discovered the intimate relationship between the real and complex parts of a complex linear functional f, namely that

$$\mathbf{Re}\, f(ix) = \mathbf{Im}\, f(x)$$

He reduced the complex case to the real case, and proved the complex version of the dominated version of the Hahn-Banach theorem for subspaces of $L_p[a,b]$ for $p > 1$. Murray's perfectly general method was used and acknowledged by Bohnenblust and Sobcyzk 1938 who proved it for arbitrary complex normed spaces. They were the first to call it the Hahn-Banach theorem. Also by reduction to the real case, Soukhomlinov 1938 and Ono 1953 obtained the theorem for vector spaces over the complex numbers and the quaternions.

Hustad 1973, Holbrook 1975 and Mira 1982 present unified approaches. Instead of reduction to the real case, each utilizes an intersection property of $\mathbf{R}$, $\mathbf{C}$, and the quaternions to prove it for all three simultaneously. We discuss these intersection properties in Sec. 6.1.

6. A Theorem for Linear Maps? (not Functionals)

Notation. Except for Secs. 10 and 11, in the sequel X and Y denote at least normed spaces over $K = \mathbf{R}$ or $\mathbf{C}$; for any vector x and $r > 0$, $B(x,r) = \{y : \|y - x\| \leq r\}$.

Can we replace the scalar field $K = \mathbf{R}$ or $\mathbf{C}$ in the Hahn-Banach theorem by a normed space Y? If $A : M \to Y$ is a bounded linear map on the closed subspace M of X, is there a linear extension $\bar{A} : X \to Y$ of A such that $\|\bar{A}\| = \|A\|$? If such an $\bar{A}$ exists for any such A on any subspace M of any normed space X, we will say that Y is **extendible**.

As we shall see, in the real case we can let Y be $\ell_\infty(n)$ for any n, even $n = \infty$—a space whose unit ball is a "cube"—but not c_0, even though its unit ball is also cubic. The extendible spaces are characterized by intersection properties of their closed balls (Theorems 6.1 and 6.2) which resemble compactness. The first one proved and the easiest to state (for the real case) is the **binary intersection property**, namely:

If $\mathcal{B}$ is a collection of mutually intersecting closed balls, then $\cap\mathcal{B} \neq \emptyset$.

As an external characterization, a space is extendible if and only if it is norm-isomorphic to some $(C(T,K), \|\cdot\|_\infty)$ where T is an extremally disconnected compact Hausdorff space

6.1. *Intersection Properties*

If Y is extendible then we must be able to continuously extend the identity map $\mathbf{1} : Y \to Y$ to $\overline{\mathbf{1}}$ on the completion $\overline{Y}$ of Y. Thus if the sequence (y_n) from Y converges to $y \in \overline{Y}$, then $\overline{\mathbf{1}}y_n = y_n \to y = \overline{\mathbf{1}}y \in Y$. Extendible spaces are therefore complete.

Since c_0 is uncomplemented in ℓ_∞ [Narici and Beckenstein 1985, p. 87], there is no continuous projection of ℓ_∞ onto c_0. This means that the identity map $y \mapsto y$ of c_0 onto c_0 does not have a continuous extension to ℓ_∞ so c_0 is not extendible. More generally, we see that an extendible space must be complemented in any Banach space in which it is norm-embedded.

As there is no loss of generality in doing so, we assume that $\|A\| \leq 1$ on M in the following discussion. The key step to extension is the ability to extend A from M to $\overline{A}$ defined on $M \oplus Kx$ for any $x \notin M$. To preserve the bound, it is necessary and sufficient to choose a value y for $\overline{A}x$ that satisfies$\|\overline{A}x - Am\| = \|y - Am\| \leq \|x - m\|$ for all $m \in M$. Thus, a permissible value y must lie in $B(Am, \|x - m\|)$ for every $m \in M$, must be in $\cap_{m\in M} B(Am, \|x-m\|)$, in other words. Since $\|y\| = \|Am\| \leq \|m\|$, we need to know what spaces Y satisfy

$$\bigcap_{m\in M} B(y, \|x-m\|) \neq \varnothing \text{ for any } y \in B(0, \|m\|) \text{ and any } x \notin M \qquad \text{(IP)}$$

By the Hahn-Banach theorem, we know that IP is satisfied in $\mathbf{R}$ and $\mathbf{C}$ no matter what M and X are, but the presence of the $M \subset X$ here is troublesome—it has nothing to do with Y. For the sake of determining purely internal characterizations of IP (Theorems 6.1 and 6.2), consider the following intersection properties.

Let S be a set and $\mathcal{B} = \{B(y_s, r_s) : y_s \in Y,\ r_s > 0,\ s \in S\}$ be a collection of closed balls in Y. If $\cap\mathcal{B} \neq \emptyset$ whenever:

- $\{B(y_s, r_s) : y_s \in Y,\ r_s > 0,\ s \in S\}$ is mutually intersecting then Y has the *binary intersection property*;
- $\bigcap\{B(f(y_s), r_s) : s \in S\} \neq \emptyset$ (in K) for any f in the unit ball of Y' then Y has the *weak intersection property*;

- for any $B(y_{s_k}, r_{s_k}) \in \mathcal{B}$, $k = 1, 2, \ldots, n$, $n \in \mathbf{N}$ and $b_1, b_2, \ldots, b_n \in K$, $\sum_{k=1}^{n} b_k = 0$ implies that $\| \sum_{k=1}^{n} b_k y_{s_k} \| \leq \sum_{k=1}^{n} |b_k| r_{s_k}$ then Y has *Holbrook's intersection property.*

Theorem 6.1. GENERAL CASE *A Banach space Y over $K = \mathbf{R}$ or $\mathbf{C}$ is extendible if and only if*

(a) [Goodner 1976] *If the Banach space X contains Y, there is a continuous projection P of norm 1 of X onto Y (Y is "1-complemented" in X);*

(b) *The identity map $\mathbf{1}: Y \to Y$ can be extended to a linear map of the same norm to any Banach space X containing Y.*

(c) *Y is topologically complemented in each space in which it is norm-embedded.*

(c) [Hustad 1973] *Y has the weak intersection property.*

(e) [Holbrook 1975, Mira 1982] *Y satisfies Holbrook's intersection property.*

(f) [Hasumi 1958] *There exists an extremally disconnected* [open sets have open closures] *compactHausdorff space T such that Y is norm-isomorphic to $(C(T, K), \|\cdot\|_\infty)$. By the Banach-Stone theorem, T is unique up to homeomorphism.*

Mira 1982 corrected an error in Holbrook's argument and also showed that if $K = \mathbf{C}$, the previous conditions are equivalent to Holbrook's intersection property being satisfied for all sets of *three* elements; if $K = \mathbf{R}$ or a non-Archimedean valued field (assuming that X has a norm which also satisfies the ultrametric triangle inequality) then we need only require it for sets of two elements, while for $K =$ the quaternions, it suffices that it be satisfied for sets of five.

Nachbin 1950 and Goodner 1950 each showed that a real extendible space *whose unit ball had an extreme point* was linearly isometric to some $C(T, \mathbf{R})$ where T is an extremally disconnected compact Hausdorff space. Nachbin had conjectured that the extreme point hypothesis was redundant. Kelley 1952 and Goodner 1976 validated his conjecture. For an extremally disconnected compact Hausdorff space T, the function $e(t) \equiv 1$ is an extreme point of the unit ball of $(C(T, \mathbf{R}), \|\cdot\|_\infty)$; hence, by Theorem 6.1(f), a necessary condition for a space to be extendible is that its unit ball possess extreme points.

Akilov 1948 provides another necessary condition for finite-dimensional spaces [cf. Goodner 1950, Cor. 4.8]: If a real finite-dimensional space Y is smooth, then it is not extendible. [A point u of the surface S of the unit ball of Y is a **smooth point** if there is a unique supporting hyperplane at u which is equivalent to Gateaux differentiability of the norm at u. Y is called **smooth**, if it is smooth at each $u \in S$.]

Theorem 6.2. REAL CASE *A real Banach space Y is extendible if and only if any of the conditions of Theorem 6.1 are satisfied as well as if and only if:*

(a) [Nachbin 1950, Goodner 1950, Kelley 1952] *Y has the binary intersection property.*

In (b) *and* (c), *the balls can be enlarged somewhat.*

(b) [Lindenstrauss 1964] *Any family $\mathcal{B} = \{B(y_s, r_s) : y_s \in Y, \ s \in S\}$ of mutually intersecting closed balls is such that for every $r > 0$, $\bigcap_{s\in S} B(y_s, (1+r)\, r_s) \neq \emptyset$.*

(c) [Davis 1977] *Any family $\mathcal{B} = \{B(y_s, 1) : y_s \in Y, \ s \in S\}$ of mutually intersecting closed unit balls is such that for every $r > 0$, $\bigcap_{s\in S} B(y_s, 1+r) \neq \emptyset$.*

(d) [Nachbin 1950, Goodner 1950, 1976, Kelley 1952] *Y is norm-isomorphic to a complete Archimedean ordered vector lattice with order unit.*

In addition to the sources cited, see also Secs. 8.8 and 10.5 of Narici and Beckenstein 1985 and Herrero 2003, pp. 149f.

Theorem 6.3. REFLEXIVITY [Goodner 1950, Theorem 6.8; Nachbin 1950, Theorem 5] *A real extendible space is reflexive if and only if it is finite-dimensional.*

For real separable spaces, we therefore have:

Theorem 6.4. SEPARABLE SPACES *Let Y be a real extendible normed space. Then Y is separable if and only if*

(a) [Goodner 1960] *Y is reflexive.*

(b) [Goodner 1950] *Y is finite-dimensional.*

(c) [Goodner 1960] *There exists a finite discrete space T such that X is norm-isomorphic to $(C(T, \mathbf{R}), \|\cdot\|_\infty)$.*

We need to define a few terms to state the next characterizations (Theorem 6.5) of extendibility of real spaces.

Definition 6.1. Let B be a bounded subset of a normed space X.

(a) The *diameter* $d(B)$ of B is $\sup\{\|x-y\| : x, y \in B\}$.

(b) The *radius* $r(B)$ of B is $\inf\{r > 0 : B \subset B(y, r),\ y \in X\}$.

In addition to boundedness, suppose that B is closed and convex for (c) and (d).

(c) B is *diametrically maximal* if for every $x \notin B$, $d(\{x\} \cup B) > d(B)$.

(d) B has *constant width* $d > 0$ if for each $f \in X'$ with $\|f\| = 1$, $\sup f(B - B) = d$.

Sets of constant width must be diametrically maximal; the two notions coincide in any two-dimensional space as well as in n-dimensional spaces with the Euclidean norm [Eggleston 1965]. They are distinct in certain three-dimensional spaces. It follows from Franchetti 1977, Moreno 2005 and Moreno, et. al. 2005 that if $Y = (C(T, \mathbf{R}), \|\cdot\|_\infty)$, where T is a compact Hausdorff space, they coincide if and only if T is extremally disconnected; this yields Theorem 6.5(b).

Theorem 6.5. RADIAL DESCRIPTIONS, REAL SPACES *A real Banach space Y is extendible if and only if any of the conditions of Theorem 6.2 are satisfied as well as if and only if:*

(a) [Davis 1977] *For every bounded subset B of Y, the diameter $d(B) = 2r(B)$.*

(b) *For every closed bounded convex subset B of Y, B has constant width if and only if B is diametrically maximal.*

6.2. *Examples on Extendible Spaces*

►As $(\mathbf{R}, |\cdot|)$ is a complete, Archimedean ordered vector lattice with order unit, it is extendible.

►When Helly 1912 proved the fundamental lemma—the one-dimensional extension—to his version of the Hahn-Banach theorem, he observed that a family of mutually intersecting closed intervals $\{[a_s, b_s] : s \in S\}$ of $\mathbf{R}$ has nonempty intersection, i.e., that $\mathbf{R}$ has the binary intersection property. He generalized this [1923] to his intersection theorem, namely that a family $\{B_s : s \in S\}$ of compact convex subsets of $\mathbf{R}^n$ has nonempty intersection if any $n+1$ of them meet; he generalized it to a topological theorem in 1930.

► $(\mathbf{R}^2, \|\cdot\|_2)$ is not extendible because it does not have the binary intersection property: There clearly exist three mutually intersecting circles whose intersection is empty. For essentially the same reason, none of $\left(\mathbf{R}^n, \|\cdot\|_p\right)$, $1 < p < \infty$, are extendible for $n > 1$; one could also argue that they are not extendible because their unit balls are smooth.

►[Nachbin 1950, Theorem 3] The only real normed spaces of finite dimension n that are extendible are those that are norm-isomorphic to $\ell_\infty(n)$.

Since the map $e_1 \mapsto e_1 + e_2$, $e_1 \mapsto e_1 - e_2$, defined on the standard basis vectors e_1 and e_2 of $\mathbf{R}^2$, is a linear isometry of real $\ell_1(2)$ onto real $\ell_\infty(2)$, it follows that real $\ell_1(2)$ is extendible.

►Real Hilbert spaces of dimension ≥ 2 do not have the binary intersection property, so are not extendible.

► $B(T, \mathbf{R})$ The real space $(B(T,\mathbf{R}), \|\cdot\|_\infty)$ of bounded real-valued functions on any set T has the binary intersection property. If $T = \{1, 2, .., n\}$ or $\mathbf{N}$ then $B(T, \mathbf{R}) = (\mathbf{R}^n, \|\cdot\|_\infty) =$ real $\ell_\infty(n)$ or real ℓ_∞, respectively.

►Since $(C([0,1], \mathbf{R}), \|\cdot\|_\infty)$ is uncomplemented in $(B([0,1], \mathbf{R}), \|\cdot\|_\infty)$, $C[0,1]$ is not extendible.

Extendibility is a geometric rather than a topological property.

►The topologies of real $\ell_p(2)$, are the same for all $1 \leq p \leq \infty$ but only $\ell_1(2)$ and $\ell_\infty(2)$ are extendible.

6.3. *The Domain*

What normed spaces X have the property that any continuous linear map A of any subspace M into any normed space Y has a linear extension with same norm?

$$\begin{array}{llll} \bar{A} : X \text{ fixed} & & & \|\bar{A}\| = \|A\| \\ \quad\;\; | \quad \searrow & & & \\ A : M & \longrightarrow & Y & A, M, Y \text{ arbitrary} \end{array}$$

If X is a Hilbert space and P_M the orthogonal projection on the closed subspace M, then $\bar{A} = A \circ P_M$ is a norm-preserving extension of A to X. And this is just about the only situation in which there are norm-preserving extensions. Kakutani 1939 (real case) and Bohnenblust 1942 and Sobczyk 1944 in the complex case (cf. also Saccoman 2001) showed that the only Banach spaces X with this property are Hilbert spaces and those X of dimension ≤ 2!

6.4. *Superspaces and Functionals*

Suppose M is a normed space over K where K is $\mathbf{R}$, $\mathbf{C}$, the quaternions or even a non-Archimedean valued field. Suppose further that for any X in which M is norm-embedded that every bounded linear *functional* on M has an extension of the same norm to X. With one proof that works for all four fields, Mira 1982 showed that these M are just the extendible spaces.

6.5. *Superspaces and Linear Maps*

Let M and Y be real Banach spaces and $A : M \to Y$ is a continuous linear map. Suppose the real Banach space X contains M as a closed subspace. Suppose further that $\bar{A}$ is a linear extension of A of the same norm to X. M is such that such extensions exist for all X, Y and A if and only if M is extendible [Nachbin 1950].

$$\begin{array}{llcl} \bar{A} : X & & & \|\bar{A}\| = \|A\| \\ \quad\ \mid & \searrow & & A,\ X,\ Y \text{ arbitrary} \\ A : M & \longrightarrow Y & & M \text{ fixed} \end{array}$$

It follows from Theorem 6.1(a), that there must be a continuous projection P of norm 1 of X onto M. Gajek *et al.* 1995 characterize such M by means of properties of the Gateaux derivative of the norm on X; for just continuous extensions, as opposed to norm-preserving ones, it suffices to be able to renorm X so that there is a continuous projection P of norm 1 of X onto M. See also Ostrovskii 2001 and Chalmers *et al.* 2003.

7. Uniqueness of the Extension

Uniqueness of continuous extensions of continuous linear functionals is closely linked to smoothness of the normed space X. For example, if M is a subspace of the normed space X and $f \in M'$ attains its norm at a smooth point, then f has a unique extension of the same norm to X. More generally, unique extensions are guaranteed for any continuous linear functional on any subspace of X if and only if X' is strictly convex (= strictly normed = rotund) which implies that X is smooth.

7.1. *Non-Uniqueness*

Consider the subspace $M = \mathbf{R} \subset \left(\mathbf{R}^2, \|\cdot\|_2\right)$ and the linear functional $f(a, 0) = a$ defined on M. Let $y \in \mathbf{R}^2$ be a unit vector of angle $\beta \neq 0, \pi$ with the x-axis. The subspace N spanned by y is a topological complement of M. For any such N, the projection P_M on M along N is a continuous extension of f and $f \circ P_M$ is a continuous extension of f of norm $|\csc \beta|$, so there are infinitely many continuous extensions of f but only one of the same norm ($\|f\| = 1$), namely when $\beta = \pm\pi/2$ (when $N = M^{\perp}$); this, incidentally, yields the extension of f of minimal norm, the smallest value of $|\csc \beta|$. Distinct extensions of f of the same norm, 1, are given by $F(a, b) = a + b$ and $G(a, b) = a - b$.

For an instance in which there are infinitely many extensions of a continuous linear functional of the same norm, consider the subspace M of constant functions of the Banach space $(C[0,1], \|\cdot\|_\infty)$ of complex-valued continuous functions on $[0,1]$ and the continuous linear functional $f : M \to \mathbf{C}$, $x \mapsto x(0)$. Clearly $\|f\| = 1$. For any $t \in [0,1]$, the evaluation map $F_t : C[0,1] \to \mathbf{C}$, $x \mapsto x(t)$ extends f and is of norm 1.

7.2. *Unique Extensions of the Same Norm: Special Cases*

Preservation of the norm clearly limits the choices that can be made for an extension and there are always extensions that do not preserve the norm.

Theorem 7.1. *If f is a continuous linear functional defined on the closed proper subspace M of the normed space X over $K = \mathbf{R}$ or $\mathbf{C}$ then there are continuous linear extensions F of f with $\|F\| > \|f\|$.*

Proof. Choose a unit vector $u \notin M$ and let d be the positive distance from u to M. Then, for any $a \in K$ and any $m \in M$, $|a|\, d \leq \|au - am\|$; indeed,

$$|a|\, d \leq \|au + m\| \quad (a \in K,\ m \in M) \tag{*}$$

Choose a scalar $b > \|f\|$. Define F on $M \oplus Ku$ by taking $F(m + au) = f(m) + ab$ for any $a \in K$ and $m \in M$. To see that F is continuous, suppose (m_n) and (a_n) are sequences from M and K, respectively, such that $m_n + a_n u \to 0$. By $(*)$, for every n, $|a_n|\, d \leq \|a_n u + m_n\| \to 0$; hence $a_n \to 0$ and $m_n \to 0$. Therefore $F(a_n u + m_n) = a_n b + f(m_n) \to 0$. Since $F(u) = b > \|f\|$, it follows that $\|F\| > \|f\|$. ■

As we argue next, extensions of the same norm are unique in Hilbert spaces.

Suppose f is a bounded linear functional defined on a closed subspace M of a Hilbert space $(X, \langle\cdot,\cdot\rangle)$, and let P_M be the orthogonal projection on M. Extend f to $F = f \circ P_M$. Since orthogonal projections are continuous, so is F. Therefore, by the Riesz representation theorem, there exist unique $m \in M$ and $n \in M^\perp$ $(X = M \oplus M^\perp)$ such that $F(\cdot) = \langle \cdot, m + n\rangle$ and $\|F\| = \|m + n\|$. Since $0 = F(n) = \|n\|^2$, it follows that $n = 0$. Hence $\|F\| = \|m\| = \|f\|$. If G is any extension of f of the same norm, a similar argument shows that $G(\cdot) = \langle \cdot, m\rangle = F(\cdot)$.

The situation for certain subspaces of the ℓ_p spaces, $1 < p < \infty$, is similar. Let $\{e_n\}$ be the standard basis for ℓ_p, let S be a subset of $\mathbf{N}$ and let f be a bounded linear functional defined on the closed linear span

M of $\{e_n : n \in S\}$. For $q = p/(p-1)$, $\|f\|^q = \sum_{n\in S} |f(e_n)|^q$. Given an extension $F((a_n)) = \sum_{n\in S} a_n f(e_n) + \sum_{n\notin S} a_n F(e_n)$ of f of the same norm, then $\|f\|^q = \|F\|^q$ implies that $F(e_n) = 0$ for all $n \notin S$.

These uniqueness results for Hilbert and ℓ_p spaces are special cases of the Taylor-Foguel theorem of Sec. 7.5.

7.3. *Uniqueness of Dominated Extensions*

Suppose f is a linear functional defined on a subspace M of a **real** vector space X and that f is dominated by a sublinear functional p (defined on X) on M. To prove that f can be extended from M to F defined on $M \oplus \mathbf{R}x_0$, $x_0 \notin M$, in such a way F is still dominated by p, a number c is chosen arbitrarily between $a = \sup\{-p(-m-x_0) - f(m) : m \in M\}$ and (the larger quantity) $b = \inf\{p(m+x_0) - f(m) : m \in M\}$ as the value for $F(x_0)$. Herein lies the non-uniqueness of F. If these two values are equal for every x, i.e., if p, f and M are such that, for each $x \in X$,

$$\sup\{-p(-m-x) - f(m) : m \in M\} = \inf\{p(m+x) - f(m) : m \in M\} \tag{2}$$

then there is only one extension F of f with $F \leq p$; conversely, if F is unique, then Eq. (2) must hold [de Guzmán 1966; cf. Herrero 2003, Th. 5.2.1]. The assertion of Eq. (2) is equivalent to, for each $x \notin M$,

$$\sup\{f(m) - p(m-x) : m \in M\} = \inf\{p(m+x) - f(m) : m \in M\}$$

Bandyopadhyay and Roy 2003 characterize when a single linear functional dominated by a sublinear functional p on a subspace M of a real vector space has a unique extension to the whole space dominated by p in terms of nested sequences of "p-balls" in a quotient space; by considering the canonical embedding of M in its bidual M'', they characterize unique extendibility of elements of M' in terms of sequences from M.

If X is **complex** [cf. Hererro 2003, Cor. 5.2.6], p a seminorm defined on X and $|f| \leq p$ on the subspace M then f has a unique extension F to X, $|F| \leq p$, if and only if for every $x \in X$,

$$\sup\{-p(-x-m) - \mathbf{Re} f(m) : m \in M\} = \inf\{p(x+m) - \mathbf{Re} f(m) : m \in M\}$$

7.4. *Unique Extensions for Points and Subspaces—Best Approximations from M^0*

Notation. M and U (or $U(X)$) denote, respectively, a closed subspace and the unit ball of a Banach space X in this section and we consider only *norm-preserving* extensions of $f \in M'$ (the continuous dual of M) to an element $F \in X'$. $L(X)$ and $K(X)$ denote, respectively, the spaces of all continuous linear operators and compact operators of X into X.

One Point If $f \in M'$ attains its norm at a smooth point, then f has a unique extension to X [Holmes 1975, p. 176].

One subspace—Phelps's theorem The seminal result characterizing subspaces M for which elements of M' have unique extensions is that of Phelps 1960: Continuous linear forms f on M have unique extensions to $f \in X'$ if and only if the annihilator $M^0 = \left\{u \in X' : u\,|_M = 0\right\}$ is Cebysev in X', in other words each $g \in X'$ has a unique best approximation $h \in M^0$, an h such that $\|g - h\| = \inf\left\{\|g - u\| : u \in M^0\right\}$.

We next consider a sufficient condition on a subspace for it to have unique extensions.

HB-subspaces M is called an *HB-subspace* if there is a projection P on X' whose kernel is M^0 and, for each $f \in X'$ such that $f \neq Pf$, $\|Pf\| < \|f\|$ and $\|f - Pf\| \leq \|f\|$. Hennefeld 1979 showed that if M is an HB-subspace, then each $f \in M'$ is uniquely extendible to X. Oja 1984 showed that a subspace may have unique extensions but not be an HB-subspace—for $X = \mathbf{R}^2$ normed by $\|(a, b)\| = \max\left(|a|, |a + b|/2\right)$, the subspace $\mathbf{R}$ has unique extensions but is not an HB-subspace. Oja 1997 gets some equivalent conditions for M to be an HB-subspace and also shows that X is an HB-subspace of its bidual whenever $K(X)$ is an HB-subspace of $L(X)$.

A subclass of the HB-subspaces for which it is easier to give examples is the M-ideals.

M-ideals M is called an *M-ideal* if there is a projection P on X' whose kernel is M^0 and, for each $f \in X'$ such that $f \neq Pf$, $\|f\| = \|f\| + \|f - Pf\|$.

Examples of M-ideals (a) If X is a B*-algebra, then any closed 2-sided ideal in X is an M-ideal [Smith and Ward 1978].

(b) Thus, spaces $(C(T), \|\cdot\|_\infty)$ of continuous functions on a compact set T are well supplied with M-ideals: For any $t \in T$, the maximal ideal $M_t = \{x \in C(T) : x(t) = 0\}$ is an M-ideal.

(c) If T is a locally compact and Hausdorff, the M-ideals of $(C_\infty(T), \|\cdot\|_\infty)$, the continuous scalar-valued functions that vanish at infinity, are precisely $M_F = \{x \in C_\infty(T) : x(F) = \{0\}\}$ where F is closed in T [Behrends 1979, p.40].

(d) For $X = L_2(\mu)$, for some measure μ, the subspace $K(X)$ of compact operators is an M-ideal in $L(X)$.

(e) The space c_0 of null sequences is an M-ideal in ℓ_∞. Thus, any continuous linear functional on c_0 has a unique Hahn-Banach extension to ℓ_∞ (Harmand *et al.* 1993, Proposition 1.12).

M-ideals and intersection properties M-ideals may be characterized internally in various ways by intersection properties of balls. For real X, the closed subspace M is an M-ideal if and only if M satisfies the *3-ball property*, namely that if three open balls B_1, B_2, B_3 have nonempty intersection and each meets M, then $M \cap \left(\cap_{i=1}^{3} B_i\right) \neq \varnothing$ [Alfsen and Effros 1972; cf. Behrends 1979, p. 46f.]. Behrends proved it for subspaces of complex spaces in 1991. Oja and Poldvere 1999 consider a related condition called the "2-ball *sequence* property" and show that M satisfies the 2-ball sequence property if and only if each $f \in M'$ has a unique extension to $F \in X'$.

Costara and Popa 2001 give further examples of subspaces for which Hahn-Banach extensions are unique.

7.5. *Unique Extensions for all Subspaces—Rotund Dual*

If the surface SU of the unit ball of X contains no nontrivial line segments (i.e., SU consists entirely of extreme points), then X is called **strictly convex** or **rotund**. Taylor 1939 proved that if the dual X' is strictly convex then any $f \in M'$ has a unique extension to $F \in X'$ of the same norm. He proved the converse when X is reflexive. Foguel 1958 removed the reflexivity, thereby showing that the normed spaces X for which each continuous linear functional on *any* subspace of X has a unique linear extension of the same norm are those X with strictly convex dual. Strict convexity of X' implies that X (not X') is smooth. Since Hilbert spaces and ℓ_p, $1 < p < \infty$, have strictly convex duals, bounded linear functionals on subspaces of either have unique extensions of the same norm. Phelps's theorem implies the Taylor-Foguel theorem (Herrero 2003, p. 88, Holmes 1975, p. 175) and was generalized by Park 1993.

The 2-ball sequence property mentioned in Sec. 7.4 provides a purely internal characterization of those X whose every Hahn-Banach extension is unique—or, equivalently, of those X with strictly convex dual—namely those X in which every closed subspace satisfies the 2-ball sequence property.

8. Non-Archimedean Functional Analysis

By considering normed spaces X over a field F with an absolute value other than **R** or **C** we can glimpse what functional analysis looks like without the Hahn-Banach theorem. There is special interest in the case when the norm and absolute value are non–Archimedean, i.e.,

$$\|x+y\| \leq \max\left(\|x\|, \|y\|\right) \text{ for all } x, y \in X$$

Even in this context, a linear functional $f : X \to F$ is continuous if and only if it is bounded on the unit ball. Non-Archimedean analysis is quite similar to ordinary analysis in situations in which the Hahn-Banach theorem holds, quite different otherwise. Because mutually intersecting balls are concentric in non-Archimedean normed spaces, the binary intersection property simplifies to:

- **Spherical Completeness.** Every decreasing sequence of closed balls has nonempty intersection.

It is similar in appearance to completeness—every decreasing sequence of closed balls *whose diameters shrink to 0* has nonempty intersection—but stronger. **R** is spherically complete. Ingleton 1952 [cf. Narici *et al.* 1977] adapted Nachbin's and Goodner's arguments about the equivalence of extendibility and the binary intersection property to prove that a non-Archimedean Banach space is extendible if and only if it is spherically complete.

Pérez-García 1992 gives a thorough survey of the Hahn-Banach extension property in the non-Archimedean case, a situation in which Hahn-Banach extensions are never unique [Beckenstein and Narici 2004]. For the case when F is not spherically complete, see Pérez-García and Schikhof 2003.

9. The Axiom of Choice

By teasing out a maximal element F from the dominating extensions of f, the standard proof of the Hahn-Banach theorem (HB) uses the Axiom of Choice (AC) in the form of Zorn's lemma.

9.1. *Is HB $\Longleftrightarrow$ AC?*

Does HB imply AC? as Tihonov's theorem does? Can we call it "the analyst's form of AC?" In a word: "No." The details are as follows.

As is well known

$$\text{AC} \implies \text{Ultrafilter theorem (UT)}$$

namely that every filter of sets is contained in an ultrafilter. Halpern 1964 proved that UT $\not\Rightarrow$ AC. Łoś and Ryll-Nardzewski 1951 and Luxemburg [1962, 1967a,b] proved that UT $\Rightarrow$ HB. Pincus [1972, 1974] proved that HB $\not\Rightarrow$ UT. We therefore have the following irreversible hierarchy:

$$\text{AC} \implies \text{UT} \implies \text{HB}$$

The "prime ideal theorem for Boolean algebras" asserts that there is a function F defined on the class of all Boolean algebras B such that $F(B)$ is a prime ideal of B for each B. Using techniques from non-standard analysis, Luxemburg 1962 showed that the prime ideal theorem implies the Hahn-Banach theorem and conjectured that the prime ideal and Hahn-Banach theorems might be equivalent. Halpern 1964, however, proved that the prime ideal theorem is strictly weaker than AC. Luxemburg 1967b showed that a modified form of the Hahn-Banach theorem is valid if and only if every Boolean algebra admits a nontrivial measure. The modification consists of allowing the extended linear functional on the real Banach space X to take values in a "reduced power of the reals" (as used in nonstandard analysis) rather than $\mathbf{R}$; the modified version is also equivalent to the unit ball of the dual of the normed space X being convex-compact in the weak-* topology, i.e., that every family of weak-*-closed convex sets with the finite intersection property has nonempty intersection. Luxemburg and Väth 2001 proved that the assertion that any Banach space has at least one nontrivial bounded linear functional implies the Hahn-Banach theorem.

9.2. *Avoiding AC*

Various people have proved weaker versions of the theorem that do not rely on the Axiom of Choice. These include:

►Garnir, de Wilde and Schmets 1968 use only the Axiom of Dependent Choices—a little stronger than the countable axiom of choice but weaker than AC—to prove a Hahn-Banach theorem for *separable* spaces.

►Ishihara 1989 proved another 'constructive' version.

►Mulvey and Pelletier 1991. *Locales* generalize the lattice of open sets of a space without reference to the points of the space. Mulvey and Pelletier avoid dependence on AC. They use locales to prove a version of the Hahn-Banach theorem in any Grothendieck topos.

►Dodu and Morillon 1999 add a little and take a little. They suppose that the Banach space X satisfies the stronger completeness requirement that Cauchy *nets* converge. They then prove the Hahn-Banach theorem for uniformly convex Banach spaces whose norm is Gateaux differentiable without AC. Still assuming that the Banach space X satisfies the stronger completeness requirement, Albius and Morillon 2001 show that to have the Hahn-Banach theorem, it suffices to have a strengthened differentiability condition, *uniform smoothness*, namely, the uniform convergence of $(\|x+h\| + \|x-h\| - 2\|x\|) / \|h\|$ as $h \to 0$ for all x on the surface of the unit ball of X.

10. "Sandwich Theorems" and Another Approach

Mazur and Orlicz 1953 used the Hahn-Banach theorem to prove interpolation theorems such as:

Theorem 10.1. *Let p be a sublinear functional and q a superlinear functional on the real vector space X such that $q \leq p$. Then there exists a linear functional f on X such that $q \leq f \leq p$.*

In a survey of results on the existence of linear functionals satisfying various conditions, Lassonde 1998 proves a blend of the Banach-Alaoglu and Hahn-Banach theorems on a real vector space to deduce results on the separation of convex functions by an affine function.

König and Rodé and others 1968–1982 reversed the direction of the usual proofs of the Hahn-Banach theorem. To describe them, let X be a real vector space and let $X^\#$ denotes the class of all sublinear (positive homogeneous, subadditive) functionals on X. Order $X^\#$ pointwise by $p \leq q$ if and only if $p(x) \leq q(x)$ for all $x \in X$. It happens that a sublinear functional p on X is linear if and only if it is a minimal element of $(X^\#, \leq)$. Given a subspace $M \subset X$ and a linear functional $f : M \to \mathbf{R}$, $f \leq p$, $p \in X^\#$, consider the collection of all sublinear functionals q such that $f \leq q$ on M and $q \leq p$ on X and choose a minimal q from this class. Any such q is a linear extension of f to X. Instead of enlarging f, they squash p. The method is considered at length in Narici and Beckenstein 1985 [Sec. 8.4].

Approaching the Hahn-Banach theorem by means of minimizing sublinear functionals leads to a variety of "sandwich theorems" such as Theorem 10.2 and its generalization below.

Theorem 10.2. *Let p be a sublinear functional defined on the real vector space X, let $S \subset X$ be convex and let $f : S \to \mathbf{R}$ be concave. If $f \leq p$ on S*

then there exists a linear functional F on X such that $f \leq F$ with $F \leq p$ on S.

Theorem 10.3. [König 1982, Th. 2.1; cf. Narici and Beckenstein 1985, Ex. 8.202(b)] *Let p be a sublinear functional defined on the real vector space X, let S be any subset of X and let $f : S \to \mathbf{R}$. If $f \leq p$ on S and there exist $a, b > 0$ such that*

$$\inf_{w \in S} \left[p\left(w - au - bv\right) - f\left(w\right) + af\left(u\right) + bf\left(v\right)\right] \leq 0 \text{ for all } u, v \in S$$

then there exists a linear functional F on X such that $f \leq F$ and $F \leq p$ on S.

Two nice surveys of this material are König 1982 and Fuchssteiner and Lusky 1981. Neumann 1994 simplifies some of the proofs of these results, develops some new ones and has a good bibliography on the subject as does Buskes 1993.

Rodé 1978 proved a very general version of the Hahn-Banach theorem, one flexible enough to apply to contexts other than linear spaces. König 1987 simplified Rodé's proof. Páles 1992 offers two geometric versions [Theorems 1 and 2] of Rodé's theorem. His Theorem 2 implies Rodé's theorem, thereby providing another proof of it.

11. Locally Convexity and Hahn-Banach Extensions

By saying that a topological vector space has the *Hahn-Banach extension property* (HBEP) we mean that any continuous linear functional on any linear subspace possesses a continuous extension to the whole space. Every locally convex Hausdorff space has HBEP. What about the converse?

In the absence of local convexity, a topological vector space X need not have any nontrivial continuous linear functionals at all. For $0 < p < 1$, for example, the dual of (the non-locally convex space) $L_p\left[0,1\right]$ is trivial [Day 1940; Kalton *et al.* 1984]. Although local convexity is not essential for the existence of nontrivial continuous linear functionals, it helps: A topological vector space X has a nontrivial dual if and only if there is a proper convex neighborhood of 0 [Köthe 1969, p. 192; Kalton *et al.* 1984. p. 17].

A. Shields had observed that, given a dual pair $\left(X, X'\right)$, any topology between the weak $\left(\sigma\left(X, X'\right)\right)$ and the Mackey topologies $\left(\tau\left(X, X'\right)\right)$ has HBEP and asked if such topologies had to be locally convex. Gregory and Shapiro 1970 showed that if $\sigma\left(X, X'\right) \neq \tau\left(X, X'\right)$ there are non-locally convex topologies in between, thereby providing a plethora of non-locally convex topologies with HBEP. Kakol 1992 gives an elementary construction for an abundance of vector topologies τ on a fixed infinite-dimensional vector space X such that (X, τ) does not have the HBEP even though X' is rich enough to separate the points of X.

Topological vector spaces can have rich duals and still not have HBEP. For $0 < p < 1$, the non-locally convex spaces ℓ_p and the Hardy spaces H_p do not have HBEP but have an abundance of continuous linear functionals such as the evaluation functionals at $n \in \mathbf{N}$ for ℓ_p or points t in the open unit disk for H_p. Indeed, $\ell_p' = \ell_\infty$ for any $0 < p < 1$ [Kalton *et al.* 1984]. Let us say that a subspace M of a TVS X has the *separation property* if any $x \notin M$ can be separated from M by a continuous linear functional. If it is possible to extend any $f \in M'$ to an element of X', we say that M has the *extension property.* For individual subspaces there is no connection between the separation and extension properties. Duren *et al.* 1969 showed that there are closed subspaces M of H^p, $0 < p < 1$, with the separation property which do not have the extension property and vice-versa. Nevertheless (*ibid.*), for an arbitrary TVS X, *every* subspace has the separation property if and only if every subspace has the extension property.

Shapiro 1970 showed that an F-space (complete metrizable TVS) X *with a basis* has the HBEP if and only if it is locally convex. Kalton removed the "with a basis" hypothesis. Using the fact that an F-space has HBEP if and only if every closed subspace is weakly closed and developing some basic sequence techniques for F-spaces, Kalton [1974; Kalton *et al.* 1984 p. 71] showed that an F-space with HBEP must be locally convex. This is false without metrizability, however—Any vector space X of uncountable algebraic dimension with the strongest vector topology (1) is not locally convex or metrizable but (2) has the Hahn-Banach extension property [Shuchat 1972].

References

1. Akilov, G. [1948] *Necessary conditions for the extension of linear operations,* C. R. (Doklady) Acad. Sci. URSS. N. S. **59**, 417–418 (Russian).
2. Albius, E. and M. Morillon [2001] *Uniform smoothness entails Hahn-Banach,* Quaest. Math. **24**, 425–439.
3. Aronszajn, N. and P. Panitchpakdi [1956], *Extensions of uniformly continuous transformations and hyperconvex metric spaces,* Pacific J. Math. **6**, 405-439.
4. Bachman, G. and L. Narici [2000] *Functional analysis,* Dover Publications,

31 East 2nd St., Mineola, NY 11501, a reprint of the 1966 edition published by Academic Press, New York.

5. Banach, S. [1923a] *Sur les opérations dans les ensembles abstraits et leur application aux équations intégrales*, Fund. Math. **3**, 133-181.
6. Banach, S. [1929] *Sur les fonctionelles linéaires*, Studia Math. **1**, 211–216 and 223–229.
7. Banach, S. [1932] *Théorie des opérations linéaires*, Chelsea, New York.
8. Banach, S. and S. Mazur [1933] *Zur Theorie der linearen Dimension*, Studia Math. **4**, 100–112.
9. Bandyopadhyay, P. and A. Roy [2003] *Nested sequences of balls, uniqueness of Hahn-Banach extensions and the Vlasov property*, Rocky Mountain J. Math. **33**, 27–67.
10. Beckenstein, E. and L. Narici [2004] *Non-uniqueness of certain Hahn-Banach extensions*, Vladikavkaz Mat. Zh. **6**, 26–28.
11. Behrends, E. [1979] *M-structure and the Banach-Stone theorem,* Springer, New York.
12. Behrends, E. [1991] *Points of symmetry of convex sets in the two-dimensional complex space—a counterexample to D. Yost's problem,* Math. Ann. **290**, 463–471.
13. Birkhoff, G. and E. Kreysig [1984] *The establishment of functional analysis*, Historia Math. **11**, 258-321.
14. Blair, Ch. E. [1977] *The Baire category theorem implies the principle of dependent choices*, Bull Acad. Polon. Sciences XXV, 933–934.
15. Bohnenblust, H. [1942] *A characterization of complex Hilbert spaces*, Port. Math. **3**, 103–109.
16. Bohnenblust, H. and A. Sobczyk [1938] *Extensions of functionals on complex linear spaces*, Bull. Amer. Math. Soc. **44**, 91–93.
17. Bombal, F. [2003] *Análisis funcional: una perspectiva histórica*, Proceedings of the Seminar of Mathematical Analysis 2002–2003, Secretariado de Publicaciones, Unversidad de Sevilla, 81–117.
18. Buskes, G. [1993] *The Hahn-Banach theorem surveyed*, Diss. Math. **327**, 1–49.
19. Butzer, P., S. Gieseler, F. Kaufmann, R. Nessel, and E. Stark [1980] *Eduard Helly (1884–1943), Eine nachträgliche Würdigung, Jahresber. Deutsch. Math.-Verein* **82**, 128–151.
20. Butzer, P., R. Nessel, and E. Stark [1984] *Eduard Helly (1884–1943), in memoriam,* Results in Mathematics **7**, 145–153.
21. Chalmers, B., M. Ostrovskii, and B. Shekhtman [2003] *Hahn-Banach operators: A review,* J. Computation Anal. Appl. **3**, 11–24.
22. Costara, C. and D. Popa [2001] *Effective determination of all the Hahn-Banach extensions of some linear and continuous functionals,* An. c Stiinc t. Univ. Ovidius Constanc ta Ser. Mat. **9**, 31–43.
23. Day, M. [1940] *The spaces* L_p *with* $0 < p < 1$, Bull. A. M. S. **46**, 816–823.
24. Davis, W. [1977] *A characterization of* P_1*-spaces*, J. Approx Theory **21**, 315–318.
25. de Guzmán, M. [1966] *Nota sobre la unicidad en el teorema de extensión*

de Hahn-Banach, Rev. Real Acad. Ciencias Exactas, Físicas y Nat. **60**, 577–584.

26. Delbaen, F. and W. Schachermayer [1994] *A general version of the fundamental theorem of asset pricing*, Math. Ann. **300**, 463–520.
27. Duren, P., B. Romberg and A. Shields [1969] *Linear functionals o H^p* spaces with $0 < p < 1$, J. Reine u. Angew. Math. **238**, 32-60.
28. Dieudonné, J. [1981a] *History of functional analysis*, North-Holland, New York.
29. Dieudonné, J. [1981b] *Choix d'œuvres mathématiques, Tome I*, 231–234, Hermann, Paris.
30. Dodu, J. and M. Morillon [1999] *The Hahn-Banach property and the axiom of choice*, MLQ Math. Log. Q. **45**, 299–314.
31. Eggleston, H. [1965] *Sets of constant width in finite-dimensional Banach spaces,* Israel J. Math. **3**, 163-172.
32. Franchetti, C. [1977] *Relationsihips between the Jung constant and a certain projection constant in Banach spaces,* Ann. Univ. Ferrara Sez. VII, N.Ser. **23**, 39–44.
33. Foguel, S. [1958] *On a theorem by A. E. Taylor,* Proc. Amer. Math. Soc. **9**, 325.
34. Fréchet, M. [1904] *Generalisation d' un théorème de Weierstrass*, C. R. Acad. Sci. **139**, 848–850.
35. Fréchet, M. [1905] *Sur les opérations linéaires*, Trans. Amer. Math. Soc. **6**, 134–140.
36. Fréchet, M. [1906] *Sur quelques points du calcul fonctionnel*, Rend. Circ. Mat. Palermo **22**, 1–74. (Fréchet's thesis.)
37. Fréchet, M. [1908] *Essai de géométrie analytique à une infinité de coordonnées*, Nouv. Ann. de Math. **8**, 97–116 and 289–317.
38. Feinberg, M., and R. Lavine [1983] *Thermodynamics based on the Hahn-Banach Theorem: The Clausius inequality,* Arch. Rat. Mech. and Anal. **82**, 203–293.
39. Fuchssteiner, B. and H. König, [1978] *New versions of the Hahn-Banach theorem*, North-Holland Math. Studies **56**, North-Holland, Amseterdam.
40. Fuchssteiner, B. and W. Lusky [1981] *Convex cones*, Proc. 2nd International Conference on General Inequalities, Math. Research Institute, Oberwolfach Black Forest, July 30–August 5, 1978, E. Beckenbach, ed., International Series of Numerical Math. **47**, Birkhäuser Verlag, Basel, 255–266.
41. Gajek, L., J. Jachymski, J., and D. Zagrodny [1995] *Projections, extendability of operators and the Gateaux derivative of the norm.* J. Appl. Anal. **1**, 29–38.
42. Garnir, H., M. de Wilde and J. Schmets [1968] *Analyse fonctionelle I*, Birkhäuser, Basel.
43. Goodner, D. [1950] *Projections in normed linear spaces*, Trans. Amer. Math. Soc. **69**, 89–108.
44. Goodner, D. [1960] *Separable spaces with the extension property*, J. London Math. Soc. **35**, 239–240.
45. Goodner, D. [1976] *Banach spaces with the bound-one projection property,*

Indian J. Math. **18**, 113–115.
46. Gregory, D. and J. Shapiro. [1970] *Nonconvex linear topologies with the Hahn-Banach extension property,* Proc. Amer. Math. Soc. **25**, 902–905.
47. Grünbaum, B. [1960] *Some applications of expanson constants,* Pacific J. Math. **10**, 193–201.
48. Hadamard, J. [1968] *Oeuvres*, 4 vols. CNRS, Paris.
49. Hahn, H. [1922] *Über Folgen linearer Operationen*, Monatsh. Math. und Phys. **32**, 1–88.
50. Hahn, H. [1927] *Über linearer Gleichungssysteme in linearer Räumen*, J. Reine Angew. Math. **157**, 214–229.
51. Halpern, J. [1964] *The independence of the axiom of choice from the Boolean prime ideal theorem*, Fund. Math. **55**, 57–66.
52. Harmand, P., D. Werner and W. Werner [1993] *M-ideals in Banach spaces and Banach algebras*, Lecture Notes in Math. 1547, Springer-Verlag, New York.
53. Hasumi, M. [1958] *The extension property of complex Banach spaces*, Tohoku Math. J. **10**, 135–142.
54. Hausdorff, F. [1914, 1917] *Grundzüge der Mengenlehre*, Veit, Leipzig.
55. Helly, E. [1912] *Über linearer Funktionaloperationen*, Österreich. Akad. Wiss. Math.-Natur. Kl. S.-B. IIa, **121**, 265–297.
56. Helly, E. [1921] *Über Systeme linearer Gleichungen mit unendlich vielen Unbekannten*, Monatsh. für Math. und Phys., **31**, 60–91.
57. Helly, E. [1923] *Über Mengen konvexer Körper mit gemeinschaftlichen Punkten,* Jahresber. Deutsch. Math.-Verein **32**, 175–176.
58. Helly, E. [1930] *Über Systeme von abgeschlossenen Mengen mit gemeinschaftlichen Punkten,* Monatsh. für Math. und Phys., **37**, 281–302.
59. Hennefeld, J. [1979] *M-ideals, HB-subspaces and compact operators,* Indiana Univ. Math. J. **28**, 927–934.
60. Herrero, P. [2003] *El teorema de Hahn-Banach: historia, versiones analíticas, geometría y aplicaciones. La propiedad de estensión*, Universidad de Murcia.
61. Holbrook, J. [1975] *Concerning the Hahn-Banach theorem*, Proc. A. M. S. **50**, 322–327.
62. Holmes, R. [1975] *Geometric functional analysis and its applications*, GTM 24, Springer-Verlag, New York.
63. Hustad, O. [1973] *A note on complex* $\mathcal{P}_1$*-spaces,* Israel J. Math. **16**, 117–119.
64. Ingleton, A. [1952] *The Hahn-Banach theorem for non-Archimedean fields*, Proc. Camb. Phil. Soc. **48**, 41–45.
65. Ioffe, A. [1981] *A new proof of the equivalence of the Hahn-Banach extension and the least upper bound properties*, Proc. A. M. S. **82**, 385–389.
66. Ishihara, H. [1989] *On the constructive Hahn-Banach theorem*, Bull. London Math. Soc. **21**, 79–81.
67. Kakol, J. [1992] *Simple construction of spaces without the Hahn-Banach extension property*, Comment. Math. Univ. Carolin. **33**, 623–624.
68. Kalton, N. [1975] *Basic sequences in F-spaces and their applications*, Proc. Edinburgh Math. Soc. **19**, 151–167.

69. Kalton, N., N. Peck and J. Roberts.[1984] *An F-space sampler*, Cambridge University Press, London Mathematical Society Lecture Notes 89, Cambridge.
70. Kakutani, S. [1939] *Some characterizations of Euclidean space*, Japanese J. Math. **16**, 93–97.
71. Kelley, J. [1952] *Banach spaces with the extension property*, Trans. Amer. Math. Soc. **72**, 323–326.
72. König, H. [1968] *Über das von Neumannsche Minimax theorem,* Arch. Math. **23**, 500–508.
73. König, H. [1970] *On certain applications of the Hahn-Banach and minimax theorems*, Arch. Math **21**, 583–591.
74. König, H. [1972] *Sublineare Funktionale,* Arch. Math **21**, 583–591.
75. König, H. [1978] *Neue Methoden und Resultate aus Funktionalanalysis und konvexer Analysis,* Oper. Res. Verf. **28**, 6–16.
76. König, H. [1980] *Der Hahn-Banach Satz von Rodé für unendlichstellige Operationen*, Arch. Math **35**, 292–304.
77. König, H. [1982] *On some basic theorems in convex analysis, in Modern applied mathematics: Optimization and operations research*, B. Korte ed., North-Holland, New York: 106–144.
78. König, H. [1987] *On the abstract Hahn-Banach theorem due to Rodé*, Aequ. Math **34**, 89–95.
79. Köthe, G. [1969] *Topological vector spaces I*, revised 1983, Springer-Verlag, New York.
80. Lassonde, M. [1998], *Uniqueness of Hahn-Banach extensions and liftings of linear dependences,* Math. Scand. **53**, 97–113.
81. Lévy, P. [1922] *Leçons d'analyse fonctionnelle*, Gauthier-Villars, Paris.
82. Lindenstrauss, J. [1964] *Extensions of compact operators,* Mem. Amer. Math. Soc. **48**.
83. Lindenstrauss, J. and L. Tzafriri [1971], *On the complemented subspace problem*, Israel J. Math. **9**, 263–269.
84. Łoś, J. and C. Ryll-Nardzewski, [1951] *On the applications of Tychonoff's theorem in mathematical proofs*, Fund. Math. **38**, 233–237.
85. Luxemburg, W. [1962] *Two applications of the method of construction by ultrapowers to analysis*, Bull. Amer. Math. Soc. **68**, 416–419.
86. Luxemburg, W. [1967a] *Beweis des Sätzes von Hahn-Banach*, Arch. Math (Basel) **18**, 271–272.
87. Luxemburg, W. [1967b] *Reduced powers of the real number system and equivalents of the Hahn-Banach theorem*, Technical Report 2, Cal. Inst. Tech. It also appeared in "Applications of Model Theory to Algebra, Analysis and Probability," Holt, Rinehart and Winston, New York, 1969, 123–137.
88. Luxemburg, W. and M. Väth [2001] *The existence of non-trivial bounded functionals implies the Hahn-Banach theorem*, Z. Anal. Anwendungen **20**, 267–279 (electronic publication).
89. Mazur, S. [1933] *Über konvexe Mengen in lineare normierte Räumen*, Studia Math. **4**, 70–84.
90. Mazur, S. and W. Orlicz [1953] *Sur les espaces métriques linéaires II,* Studia

Math. **13**, 137–179.

91. Minkowski, H. [1896] *Geometrie der Zahlen*, Teubner, Leipzig.
92. Mira, J. [1982] *A unified approach to the extension problem for normed spaces*, Boll. Un. Mat. Ital. **1**, 225–232.
93. Moreno, J. [2005] *Semicontinuous functions and convex sets in C(K) spaces*, J. Australian Math. Soc., to appear.
94. Moreno, J., P. Papini and R. Phelps [2005] *Diametrically maximal and constant width sets in Banach spaces*, Canadian J. Math., to appear.
95. Mulvey C. and J. Pelletier [1991] *A globalization of the Hahn-Banach Theorem*, Adv. Math. **89**, 1–59.
96. Murray, F. [1936] *Linear transformations in L_p, $p > 1$*, Trans. Amer. Math. Soc. **39**, 83–100.
97. Murray, F. [1937] *On complementary manifolds in spaces L_p and ℓ_p*, Trans. Amer. Math. Soc. **41**, 138–152.
98. Nachbin, L. [1950] *A theorem of Hahn-Banach type for linear transformations*, Trans. Amer. Math. Soc. **68**, 28–46.
99. Narici, L., E. Beckenstein and G. Bachman [1971] *Functional analysis and valuation theory*, Marcel Dekker, New York.
100. Narici, L. and E. Beckenstein [1985] *Topological vector spaces*, Marcel Dekker, New York.
101. Narici, L. and E. Beckenstein [1997] *The Hahn-Banach theorem: the life and times*, Topology and its Applications **77**, 193–211.
102. Neumann, M. [1994] *Sandwich theorems for operators of convex tyupe*, J. Math. Anal. and Appl. **188**, 759–773.
103. Neumann, M. and V. Velasco [1998] *Network analysis*, Minimax Theory and Applications, ed. by B. Ricceri and S. Simons, Kluwer, Dordrecht,167–189.
104. Oja, E. [1984] *On the uniqueness of the norm presrving extension of a linear functional in the Hahn-Banach theorem,* Iav. Akad. Nauk Es. SSR **33**, 424–439 (Russian).
105. Oja, E. [1997] *HB-subspaces and Godun sets of subspaces in Banach spaces,* Mathematika **44**, 120–132.
106. Oja, E. and M. Poldvere [1999] *Intersection of ball sequences and uniqueness of Hahn-Banach extensions,* Proc. Royal Soc. Edinburgh **129A**, 1251-1262.
107. Ono, T. [1953] *A generalization of the Hahn-Banach theorem*, Nagoya Math. J. **6**, 171–176.
108. Ostrovskii, M. [2001] *Hahn-Banach operators*, Proc. Amer. Math. Soc., electronic publication.
109. Páles, Z. [1992] *Geometric versions of Rodé's theorem,* Rad. Mat. **8**, 217–229.
110. Park, S. [1993] *A little generalization of the Hahn-Banach extension property*, J. Korean Math. Soc. **30**, 139–150.
111. Peano, G. [1888] *Calcolo geometrico secondo l'Ausdehnungslehre di H. Grassmann preceduto dalle operazioni della logica deduttiva*, Bocca, Turin; see also the reproduction in Formulario Matematico, vol. V, Edizioni Cremonese, Rome, 1960.
112. Pérez-García, C. [1992] *The Hahn-Banach extension property in p-adic*

analysis, in *P-adic functional analysis*, Lecture Notes in Pure and Appl. Math. **137,** Marcel Dekker, New York, 127–140.

113. Pérez-García, C. and W. Schikhof [2003] *Finite-dimensional orthocomplemented subspaces in p-adic normed spaces.* Ultrametric functional analysis, Contemp. Math. **319**, Amer. Math. Soc., Providence, RI, 281–298.
114. Phelps, R. [1960] *Uniqueness of Hahn-Banach extensions and unique best approximation,* Trans. Amer. Math. Soc. **95**, 238–255.
115. Pincherle, S. [1954] *Opere scelte*, 2 vol., Cremonese, Roma.
116. Pincus, D., [1972] *Independence of the prime ideal theorem from the Hahn-Banach theorem,* Bull. Amer. Math. Soc. **78**, 766–770.
117. Pincus, D. [1974] *The strength of the Hahn-Banach theorem,* Proc. Victoria Symposium on Nonstandard Analysis, Springer-Verlag, New York, 203–208.
118. Riesz, F. [1910] *Sur certain systèmes d'équations fonctionelles et l'approximation des fonctions continues,* Académie des Sciences, Paris, Comptes Rendus **150**, 674–677.
119. Riesz, F. [1911] *Sur certain systèmes singuliers d'équations intégrales*, Ann. Sci. École Norm. Sup. **28**, 33–62.
120. Riesz, F. [1913] *Les systèmes d'équations linéaires à une infinité d'inconnues*, Gauthier-Villars, Paris.
121. Riesz, F. [1918] *Über lineare Funktionalgleichungen*, Acta Math. **41**, 71-98. Also in his complete works (1960 below) 1053–1080.
122. Riesz, F. [1960] *Oeuvres complètes*, 2 vol., Akadémiai Kiadó, Budapest.
123. Rodé, G. [1978] *Eine abstrakte Versiion des Sätzes von Hahn-Banach,* Arch. Math. (Basel) **31** (1978/79), 474–481.
124. Saccoman, J. [2001] *On the extension of linear operators*, Int. J. Math. Math. Sci. **28**, 621–623.
125. Schmidt, E. [1908] *Über die Auflösung linearer Gleichungen mit unendlich vielen Unbekannten*, Rend. Palermo XXV, 53–77.
126. Shapiro, J. [1970] *Existence of linear functionals on F-spaces with bases,* Duke Math. J. **37**, 639–645.
127. Shuchat, A. [1972] *Approximation of vector-valued continuous functions,* Proc. A. M. S. **31**, 97–103.
128. Simons, S. [1970a] *Minimal sublinear functionals*, Studia Math. **37**, 37–56.
129. Simons, S. [1970b] *Formes souslinéaires minimales*, Sém. Choquet 1970/71, no. 23.
130. Simons, S. [1975] *Convergence theorems, Hahn-Banach and Choquet theorems, minimax theorem and James's theorem, in Analyse fonctionelle et applications*, L. Nachbin, ed., Hermann, Paris, 271–276.
131. Simons, S. [2003] *A new version of the Hahn-Banach theorem*, Arch. Math. (Basel) **80**, 630–646.
132. Smith, R. and J. Ward [1978] *M-ideal structure in Banach algebras*, J. Func. Anal. **27**, 337–349.
133. Sobczyk, A. [1944] *On the extension of linear transformations*, Trans. A.M.S. **55**, 153–169.
134. Soukhomlinov, G. A. [1938] *On the extension of linear functionals in complex and quaternion linear spaces*, Matem. Sbornik **3**, 353–358 [Russian with

German summary].
135. Taylor, A. [1939] *The extension of linear functionals,* Duke Math. J. **5**, 538–547.
136. Volterra, V., [1887] *Sopra le funzioni che dipendono da altre funzioni,* Rend. Accad. Lincei **3**, 97–105; see also *Opere matematiche*, 5 vol., Acc. dei Lincei, 294–302, 1954–1962.
137. Weber, H. *Lehrbuch der algebra*, 2nd ed., 3 vol., Braunschweig, Vieweg, 1898–1908.
138. Wiener, N. [1922] *Limit in terms of continuous transformation*, Bull. Soc. Math. Fr. **50**, 119–134.
139. Wiener, N. [1923] *Note on a paper of M. Banach*, Fund. Math. **4**, 136–143.
140. Zippin, M. [2003] *Extensions of bounded linear operators*, Handbook of the Geometry of Banach Spaces, vol. 2, 1703–1741,

SPECTRAL PROPERTIES OF CESÀRO-LIKE OPERATORS

MICHAEL M. NEUMANN

Abstract The spectral picture, the question of hyponormality, and other spectral decomposition properties are discussed for the class of generalized Cesàro operators with rational symbol on Hardy and Bergman spaces.[a]

1. Introduction and outline

The purpose of this article is to survey some of the recent developments in the spectral theory of certain averaging operators on Banach spaces of analytic functions on the unit disc. Our main interest will be in generalized Cesàro operators with rational symbol on Hardy spaces, but we shall also consider the case of weighted Bergman spaces. We shall focus on two main streams of investigation for such operators.

On the one hand, a natural point of central interest is the spectral picture of such operators in terms of the symbol. In this direction, we shall obtain a precise description of the spectrum, the point spectrum, the essential spectrum, and the index. This line of research dates back to classical work by G. H. Hardy [11], A. Brown, P. R. Halmos and A. L. Shields [7], and M. González [10]. Here we build on ideas and results mainly due to A. G. Siskakis [23], [24], [25], [26] and S. W. Young [27], [28], [29], and present the state of the art from the recent collaboration of E. Albrecht, T. L. Miller, and the author [2]. One of the fascinations of the subject lies in the powerful use of Banach algebra techniques in this branch of function-theoretic operator theory. More precisely, for a generalized Cesàro operator with rational symbol, it turns out to be most convenient to work in the Calkin algebra and thus to obtain a formula for the essential spectrum and the index before addressing the spectrum itself.

---

[a]2000 *Mathematics Subject Classification.* Primary 47B38; Secondary 46E15, 47A10, 47B40.

Keywords and phrases: Cesàro operator, Fredholm operator, index, subnormality, hyponormality, decomposable operator, Bishop's property (β).

Our second emphasis will be on spectral decomposition properties of generalized Cesàro operators with rational symbol, with emphasis on the question of subnormality, hyponormality, or subdecomposability in the sense of C. Foiaş. This line of research was initiated by the seminal work of T. L. Kriete and D. Trutt [14], who established, in a remarkable tour de force, the subnormality of the classical Cesàro operator after hyponormality was verified in [7]. Here we extend recent results on hyponormality for generalized Cesàro operators from [2] and discuss subdecomposability in the spirit of work due to T. L. Miller, V. G. Miller, R. C. Smith, and the author [16], [17], [18], [19], [20].

This article is an expanded version of a plenary talk presented at the Second International Course of Mathematical Analysis in Andalucía at the University of Granada in September 2004. The warm atmosphere of this conference, the excellent organization, and the generous hospitality of the organizers are gratefully acknowledged.

2. Basic definitions and notation

Throughout this note, we shall use the standard notation of operator theory and local spectral theory as employed, for instance, in [15]. In particular, $\mathcal{L}(X)$ stands for the Banach algebra of all bounded linear operators on a complex Banach space X, and, for arbitrary $T \in \mathcal{L}(X)$, the *spectrum* and *point spectrum* of T are denoted by $\sigma(T)$ and $\sigma_{\mathrm{p}}(T)$, respectively. T is said to be a *Fredholm operator* if its kernel, $\ker(T)$, is finite dimensional and its range, $\mathrm{ran}(T)$, is of finite codimension in X. The *index* of a Fredholm operator $T \in \mathcal{L}(X)$ is the integer $\mathrm{ind}(T) := \dim\ker(T) - \dim X/\mathrm{ran}(T)$. The *essential spectrum* of an arbitrary operator $T \in \mathcal{L}(X)$ is defined as the set $\sigma_{\mathrm{e}}(T)$ of all $\lambda \in \mathbb{C}$ for which the operator $\lambda I - T$ fails to be a Fredholm operator, where I stands for the identity operator on X. It is well known that $\sigma_{\mathrm{e}}(T)$ coincides with the spectrum of T in the Calkin algebra $\mathcal{L}(X)/\mathcal{K}(X)$, where $\mathcal{K}(X)$ denotes the ideal of compact operators on X.

For a complex Hilbert space X, recall that an operator $T \in \mathcal{L}(X)$ is *self-adjoint* if $T = T^*$, and that T is *normal* if $T^*T = TT^*$, equivalently, if $\|T^*x\| = \|Tx\|$ for all $x \in X$. Moreover, T is *subnormal* if T is the restriction of some normal operator to a closed invariant subspace, and T is *hyponormal* if $\langle (T^*T - TT^*)\, x, x\rangle \geq 0$ for all $x \in X$, equivalently, if $\|T^*x\| \leq \|Tx\|$ for all $x \in X$. Each of the last three conditions is weaker than the preceding one.

Also, an operator $T \in \mathcal{L}(X)$ on a complex Banach space X is said to be *decomposable* in the sense of C. Foiaş if, for each open cover $\{U, V\}$ of $\mathbb{C}$, there exist T-invariant closed linear subspaces Y and Z of X for which $X = Y + Z$, $\sigma(T \mid Y) \subseteq U$, and $\sigma(T \mid Z) \subseteq V$. Simple examples of decomposable operators include the normal, compact, and Riesz operators as well as all operators with a continuous C^∞-functional calculus. Also, all multiplication operators on a regular semi-simple commutative Banach algebra are decomposable, and so are many (but not all) convolution operators on the group algebra $L^1(G)$ and the measure algebra $M(G)$ for a locally compact abelian group G; see [15] for details and further information.

Finally, an operator is said to be *subdecomposable* if it is the restriction of some decomposable operator to a closed invariant subspace. As shown by E. Albrecht and J. Eschmeier [1], subdecomposability may be intrinsically characterized by a classical condition introduced by E. Bishop that has now become one of the central notions of modern local spectral theory. Clearly, all subnormal operators are subdecomposable, and, by a seminal result due to M. Putinar [22], the same is true for all hyponormal operators. However, as will be seen below, there are many operators of Cesàro-type which are subdecomposable, but fail to be hyponormal.

3. The classical Cesàro operator

The classical Cesàro operator is the mapping that assigns to each square-summable sequence of complex numbers the corresponding sequence of Cesàro means, so that

$$C(x) := \left(\frac{x_0 + x_1 + \cdots + x_n}{n+1} \right)_{n \in \mathbb{N}_0} \qquad \text{for all } x \in \ell^2(\mathbb{N}_0)$$

where $\mathbb{N}_0 := \{0, 1, 2, \ldots\}$. In 1920, G. H. Hardy [11] realized that it is possible to provide a very short and elementary proof of a classical theorem on double series due to D. Hilbert, once it is known that the Cesàro operator maps the Hilbert space $\ell^2(\mathbb{N}_0)$ into itself. For this Hardy presented a beautiful argument that was communicated to him by M. Riesz: for a non-negative sequence $x \in \ell^2(\mathbb{N}_0)$, let

$$s_k := \sum_{j=0}^{k} x_j \qquad \text{and} \qquad r_k := \sum_{j=k+1}^{\infty} \frac{1}{j^2},$$

and observe that $s_k^2 - s_{k-1}^2 \leq 2\, x_k\, s_k$ and

$$r_k \leq \frac{1}{(k+1)^2} + \int_{k+1}^{\infty} \frac{\mathrm{d}x}{x^2} = \frac{1}{(k+1)^2} + \frac{1}{k+1} \leq \frac{2}{k+1}$$

for all $k \in \mathbb{N}_0$ with $s_{-1} := 0$. From this and the Cauchy–Schwarz inequality it follows that

$$\sum_{k=0}^{n} \left(\frac{s_k}{k+1} \right)^2 = \sum_{k=0}^{n} s_k^2\,(r_k - r_{k+1}) = \sum_{k=0}^{n} \left(s_k^2 - s_{k-1}^2\right) r_k - s_n^2\, r_{n+1}$$
$$\leq 4 \sum_{k=0}^{n} x_k \frac{s_k}{k+1} \leq 4 \left(\sum_{k=0}^{n} x_k^2 \right)^{1/2} \left(\sum_{k=0}^{n} \left(\frac{s_k}{k+1} \right)^2 \right)^{1/2}$$

for all $n \in \mathbb{N}_0$ and therefore $\|C(x)\|_2 \leq 4\,\|x\|_2$, first for all non-negative $x \in \ell^2(\mathbb{N}_0)$ and then, of course, for arbitrary $x \in \ell^2(\mathbb{N}_0)$. This argument marks the birth of the Cesàro operator as a bounded linear operator on $\ell^2(\mathbb{N}_0)$. In the preceding estimate, the constant 4 is not optimal, and a number of alternative approaches were later developed to establish that the Cesàro operator is actually bounded on $\ell^p(\mathbb{N}_0)$ with norm $p/(p-1)$ for each $p > 1$, a fact that has become known as Hardy's inequality; see Theorem 326 of the book by G. H. Hardy, J. E. Littlewood, and G. Pólya [12].

The systematic investigation of the spectral properties of the Cesàro operator on the Hilbert space $\ell^2(\mathbb{N}_0)$ was initiated by A. Brown, P. R. Halmos, and A. L. Shields [7] in 1965. They proved that the spectrum $\sigma(C)$ is the closed disc $\overline{D(1)}$, where we use the notation

$$D(a) := \{\lambda \in \mathbb{C} : |\lambda - a| < |a|\} \qquad \text{for non-zero } a \in \mathbb{C}.$$

Moreover, the point spectrum $\sigma_\mathrm{p}(C)$ of C is empty, while $\sigma_\mathrm{p}(C^*) = D(1)$, and all eigenvalues of C^* are simple. It was also shown in [7] that C is hyponormal. In 1971, this result was then improved by T. L. Kriete and D. Trutt [14], who established that C is, in fact, subnormal by constructing an explicit normal extension of C. The spectral picture of C was completed in 1985 by M. González [10], who proved that the essential spectrum of C is the boundary $\partial D(1)$ of its spectrum and that the index of C is constant $= -1$ on the interior $D(1)$ of the spectrum.

To obtain another view of the Cesàro operator that opens the door to useful extensions, we perform a metamorphosis into function-theoretic operator theory. As usual, let $\mathbb{D} := \{\lambda \in \mathbb{C} : |\lambda| < 1\}$ be the open unit disc, and let $H(\mathbb{D})$ denote the Fréchet space of all analytic complex-valued functions on $\mathbb{D}$, endowed with the topology of locally uniform convergence. It is well known that the space $H^2(\mathbb{D})$ consisting of all $f \in H(\mathbb{D})$ for which

$$\|f\|^2 := \sup_{r<1} \frac{1}{2\pi} \int_0^{2\pi} \left| f\left(r\,\mathrm{e}^{\mathrm{i}\theta}\right)\right|^2 \mathrm{d}\theta < \infty,$$

is, with the canonical norm, a Hilbert space with orthonormal basis $(z^n)_{n\in\mathbb{N}_0}$. Moreover, there is an isometric isomorphism $H^2(\mathbb{D}) \cong \ell^2(\mathbb{N}_0)$ via the correspondence given by

$$\sum_{k=0}^{\infty} c_k\, z^k \leftrightarrow (c_0, c_1, c_2, \ldots).$$

A similar definition holds for $H^p(\mathbb{D})$ for $1 \le p < \infty$, while $H^\infty(\mathbb{D})$ denotes the Banach algebra of all bounded analytic functions on $\mathbb{D}$. For an excellent exposition of the theory of the Hardy spaces $H^p(\mathbb{D})$ for $1 \le p \le \infty$, we refer to P. Duren [9].

In terms of the canonical identification of the spaces $H^2(\mathbb{D})$ and $\ell^2(\mathbb{N}_0)$, we obtain a convenient integral representation of the Cesàro operator, namely

$$(Cf)(z) = \frac{1}{z}\int_0^z \frac{f(\omega)}{1-\omega}\,\mathrm{d}\omega \qquad \text{for all } f \in H^2(\mathbb{D}) \text{ and } z \in \mathbb{D},$$

with the obvious interpretation for $z = 0$. Indeed, using the geometric series and the power series representation $f(z) = \sum\limits_{k=0}^{\infty} c_k\, z^k$ of a function $f \in H^2(\mathbb{D})$, we arrive at

$$\frac{1}{z}\int_0^z \frac{f(\omega)}{1-\omega}\,\mathrm{d}\omega = \frac{1}{z}\int_0^z \left(\sum_{j=0}^{\infty} \omega^j\right)\left(\sum_{k=0}^{\infty} c_k\,\omega^k\right)\mathrm{d}\omega$$

$$= \frac{1}{z}\int_0^z \sum_{n=0}^{\infty} (c_0 + c_1 + \cdots + c_n)\,\omega^n\,\mathrm{d}\omega = \sum_{n=0}^{\infty} \frac{c_0 + \cdots + c_n}{n+1}\, z^n.$$

The preceding integral representation motivates the class of operators to be considered in the next section.

4. Generalized Cesàro operators

For arbitrary $g \in H(\mathbb{D})$, let

$$(S_g f)(z) := \frac{1}{z}\int_0^z f(\omega) g(\omega)\,\mathrm{d}\omega \qquad \text{for all } f \in H(\mathbb{D}) \text{ and } z \in \mathbb{D}.$$

Evidently, S_g is linear and continuous on $H(\mathbb{D})$. This operator is often called a generalized Cesàro operator with symbol g. The following observation is straightforward.

Lemma 4.1. *If* $g(z) = \sum_{k=0}^{\infty} c_k z^k$ *induces a bounded operator* S_g *on* $H^2(\mathbb{D})$, *then the matrix representation of* S_g *on* $H^2(\mathbb{D})$ *is*

$$[S_g] = \begin{bmatrix} c_0 & 0 & 0 & 0 & 0 & \dots \\ \frac{c_1}{2} & \frac{c_0}{2} & 0 & 0 & 0 & \dots \\ \frac{c_2}{3} & \frac{c_1}{3} & \frac{c_0}{3} & 0 & 0 & \dots \\ \frac{c_3}{4} & \frac{c_2}{4} & \frac{c_1}{4} & \frac{c_0}{4} & 0 & \dots \\ \frac{c_4}{5} & \frac{c_3}{5} & \frac{c_2}{5} & \frac{c_1}{5} & \frac{c_0}{5} & \dots \\ \vdots & \vdots & \vdots & \vdots & \vdots & \ddots \end{bmatrix}.$$

Clearly, for the classical Cesàro operator, $c_n = 1$ for all $n \geq 0$, again by the formula for the geometric series. An immediate consequence of the preceding result is that generalized Cesàro operators are rarely normal or self-adjoint. More precisely, we have:

Corollary 4.1. S_g *is normal on* $H^2(\mathbb{D})$ *precisely when* $g(z) \equiv c_0 \in \mathbb{C}$, *while* S_g *is self-adjoint on* $H^2(\mathbb{D})$ *precisely when* $g(z) \equiv c_0 \in \mathbb{R}$.

Proof This is clear from a computation of the matrix entries in row 0 and column 0 for $S_g^* S_g$ and $S_g S_g^*$, namely $\left[S_g^* S_g\right]_{0,0} = \sum_{k=0}^{\infty} \frac{|c_k|^2}{(k+1)^2}$ and $\left[S_g S_g^*\right]_{0,0} = |c_0|^2$. □

The problem of continuity and compactness of generalized Cesàro operators on $H^p(\mathbb{D})$ was settled by A. Aleman and A. G. Siskakis [5] for arbitrary $1 \leq p < \infty$ after classical work of Ch. Pommerenke [21] on the case $p = 2$. A. Aleman and J. A. Cima [4] obtained similar results for $0 < p < 1$.

Theorem 4.1. *Let* $1 \leq p < \infty$, *and put* $G(z) := \int_0^z g(\omega)\, d\omega$ *for all* $z \in \mathbb{D}$. *Then*

(a) S_g *is bounded on* $H^p(\mathbb{D}) \iff G$ *is of bounded mean oscillation;*
(b) S_g *is compact on* $H^p(\mathbb{D}) \iff G$ *is of vanishing mean oscillation.*

In the following, we do not require any knowledge of functions of bounded mean oscillation, since we shall restrict our attention to a special class of symbols.

Let $1 < p < \infty$ be given. It is elementary to check that S_g is compact on $H^p(\mathbb{D})$ whenever $g \in H^\infty(\mathbb{D})$; see Corollary 5.1 below. Also, for arbitrary $b \in \partial\mathbb{D}$, let

$$(C_b f)(z) := \frac{1}{z}\int_0^z \frac{f(\omega)}{1-b\,\omega}\,\mathrm{d}\omega \qquad \text{for all } f \in H^p(\mathbb{D}) \text{ and } z \in \mathbb{D}.$$

Thus C_1 is the Cesàro operator on $H^p(\mathbb{D})$ that was studied by A. G. Siskakis [23].

Lemma 4.2. *C_1 is bounded on $H^p(\mathbb{D})$ with $\sigma_{\mathrm{p}}(C_1) = \emptyset$, $\sigma(C_1) = \overline{D(p/2)}$, $\sigma_{\mathrm{e}}(C_1) = \partial D(p/2)$, and* $\operatorname{ind}(C_1 - \lambda I) = -1$ *for all $\lambda \in D(p/2)$.*

We observe that, for arbitrary $b \in \partial\mathbb{D}$, the operator C_b is similar to C_1, since the composition operator U_b given by $(U_b f)(z) := f(bz)$ is a surjective isometry on $H^p(\mathbb{D})$ with the property that $C_b = U_b C_1 U_{\overline{b}}$.

In particular, it follows that S_g is bounded on $H^p(\mathbb{D})$ for any function of the form

$$g(z) = \sum_{j=1}^{m} \frac{a_j}{1-b_j z} + h(z) \qquad \text{for all } z \in \mathbb{D},$$

where $b_1, \ldots, b_m \in \partial\mathbb{D}$ are distinct, $a_1, \ldots, a_m \in \mathbb{C} \setminus \{0\}$, and $h \in H^\infty(\mathbb{D})$. In the special case $p = 2$, such generalized Cesàro operators with rational symbol were investigated by S. W. Young [27], [28] using Hilbert space methods. In the next section, we shall discuss the fine spectrum of such operators for arbitrary $1 < p < \infty$ in the spirit of [2]. Note that, if a rational function g induces a bounded operator on $H^p(\mathbb{D})$, then necessarily g has no poles in $\mathbb{D}$ and only finitely many poles of multiplicity 1 on $\partial\mathbb{D}$. In fact, it is not difficult to see that S_g is not defined on the entire space $H^p(\mathbb{D})$, if $g(z) = (1-bz)^{-2}$ and $|b| = 1$.

5. The spectral picture

The next results are the main tools from [2] to compute the spectrum of generalized Cesàro operators with rational symbol. Basically, Lemma 5.1 follows from the fact that, for any $0 < t < 1$, the dilation operator given by $(U_t f)(z) := f(tz)$ is compact on $H^p(\mathbb{D})$.

Lemma 5.1. *Let $\varphi \in L^1([0,1])$, and define K_φ on $H(\mathbb{D})$ by*

$$(K_\varphi f)(z) := \int_0^1 f(t\,z)\varphi(t)\,\mathrm{d}t \qquad \textit{for all } f \in H(\mathbb{D}) \textit{ and } z \in \mathbb{D}.$$

Then K_φ is compact on $H^p(\mathbb{D})$ for each $1 \le p < \infty$.

Corollary 5.1. *If $g \in H^\infty(\mathbb{D})$, then S_g compact on $H^p(\mathbb{D})$.*

Proof Note that $S_g = K_1 M_g$, where M_g denotes multiplication by g. □

Corollary 5.2. *If $b_1, b_2 \in \partial\mathbb{D}$ are distinct, then $C_{b_1} C_{b_2}$ is compact on $H^p(\mathbb{D})$.*

Proof A simple computation based on a suitable partial fraction decomposition and Fubini's theorem yields that $C_{b_1} C_{b_2} = b_2 K_{\varphi_2} C_{b_2} - b_1 C_{b_1} K_{\varphi_1}$, where

$$\varphi_1(s) := \frac{1}{b_2 s - b_1} \quad \text{and} \quad \varphi_2(s) := \frac{s}{b_2 s - b_1}$$

are integrable on $[0,1]$. Thus Lemma 5.1 applies. □

Now, let $1 < p < \infty$, and, for distinct $b_1, \ldots, b_m \in \partial\mathbb{D}$, non-zero $a_1, \ldots, a_m \in \mathbb{C}$, and $h \in H^\infty(\mathbb{D})$, consider the operator S_g on $H^p(\mathbb{D})$ given by

$$g(z) := \sum_{j=1}^{m} \frac{a_j}{1 - b_j z} + h(z) \qquad \text{for all } z \in \mathbb{D},$$

and define $D_j := D\left(a_j p/2\right)$ for $j = 1, \ldots, m$.

Theorem 5.1. *$\sigma_{\mathrm{e}}(S_g) = \bigcup_{j=1}^{m} \partial D_j$ and $\operatorname{ind}(\lambda I - S_g) = -\sum_{j=1}^{m} \chi_{D_j}(\lambda)$ for all $\lambda \in \mathbb{C} \setminus \sigma_{\mathrm{e}}(S_g)$.*

Proof By Corollary 5.1,

$$\sigma_{\mathrm{e}}(S_g) = \sigma_{\mathrm{e}} \left(\sum_{j=1}^{m} a_j C_{b_j} \right),$$

and, by similarity and Lemma 4.2, $\sigma_{\mathrm{e}}\left(a_j C_{b_j}\right) = a_j \sigma_{\mathrm{e}}\left(C_{b_j}\right) = a_j \sigma_{\mathrm{e}}(C_1) = a_j \partial D(p/2) = \partial D_j$. Also, for non-zero $\lambda \in \mathbb{C}$,

$$\prod_{j=1}^{m} \left(\lambda I - a_j C_{b_j}\right) = \lambda^m I - \lambda^{m-1} \sum_{j=1}^{m} a_j C_{b_j} + H,$$

where H is compact by Corollary 5.2. Thus

$$\lambda I - \sum_{j=1}^{m} a_j C_{b_j} = \lambda^{1-m} \prod_{j=1}^{m} \left(\lambda I - a_j C_{b_j}\right) - \lambda^{1-m} H,$$

and consequently

$$\lambda \notin \sigma_{\mathrm{e}}\left(\sum_{j=1}^{m} a_j\, C_{b_j}\right) \Longleftrightarrow \lambda \notin \sigma_{\mathrm{e}}\left(a_j\, C_{b_j}\right) \text{ for } j=1,\ldots,m.$$

This implies that

$$\sigma_{\mathrm{e}}(S_g) = \sigma_{\mathrm{e}}\left(\sum_{j=1}^{m} a_j\, C_{b_j}\right) = \bigcup_{j=1}^{m} \sigma_{\mathrm{e}}\left(a_j\, C_{b_j}\right) = \bigcup_{j=1}^{m} \partial D_j.$$

Moreover, for $\lambda \in \mathbb{C} \setminus \sigma_{\mathrm{e}}(S_g)$, the index theorem ensures that

$$\begin{aligned}\operatorname{ind}(\lambda I - S_g) &= \operatorname{ind}\left(\lambda I - \sum_{j=1}^{m} a_j\, C_{b_j}\right) = \operatorname{ind}\left(\prod_{j=1}^{m}\left(\lambda I - a_j\, C_{b_j}\right)\right)\\ &= \sum_{j=1}^{m} \operatorname{ind}\left(\lambda I - a_j\, C_{b_j}\right) = -\sum_{j=1}^{m} \chi_{D_j}(\lambda),\end{aligned}$$

as claimed. □

Theorem 5.2. *If $g(0)=0$, then $\sigma_{\mathrm{p}}(S_g)=\emptyset$. Otherwise*

$$\sigma_{\mathrm{p}}(S_g) = \left\{\frac{g(0)}{k} : k \in \mathbb{N} \textit{ with } \operatorname{Re}\left(\frac{a_j}{g(0)}\right) < \frac{1}{k\,p} \textit{ for } j=1,\ldots,m\right\},$$

and each eigenvalue of S_g is simple.

Proof Suppose that $\lambda \neq 0$ and $0 \neq f \in H(\mathbb{D})$ such that $S_g f = \lambda f$. Writing $f(z) = z^n \varphi(z)$ with $n \geq 0$ and $\varphi(0) \neq 0$, we obtain that

$$\int_0^z \omega^n \varphi(\omega)\, g(\omega)\, \mathrm{d}\omega = \lambda\, z^{n+1} \varphi(z) \qquad \text{for all } z \in \mathbb{D}.$$

Differentiation then yields

$$\varphi(z)\, g(z) = \lambda\,(n+1)\,\varphi(z) + \lambda\, z\, \varphi'(z) \qquad \text{for all } z \in \mathbb{D}. \qquad (\diamondsuit)$$

Evaluation at $z=0$ entails $g(0) = \lambda\,(n+1)$, and therefore $\sigma_{\mathrm{p}}(S_g) \subseteq \{g(0)/k : k \in \mathbb{N}\}$. Also, $\sigma_{\mathrm{p}}(S_g) = \emptyset$ if $g(0)=0$, since $\lambda = 0$ cannot be an eigenvalue. The stipulated formula for $\sigma_{\mathrm{p}}(S_g)$ now follows from an elementary explicit solution of the differential equation $(\diamondsuit)$ and an investigation of the condition when this solution belongs to $H^p(\mathbb{D})$, based on the fact that $(1-z)^\mu \in H^p(\mathbb{D})$ precisely when $\operatorname{Re}(\mu) > -1/p$. □

Evidently, the preceding two results allow us to compute the spectrum as follows.

Theorem 5.3. $\sigma(S_g) = \sigma_{\mathrm{p}}(S_g) \cup \bigcup_{j=1}^{m} \overline{D(a_j p/2)}$.

As observed in [2], $\sigma_{\mathrm{p}}(S_g)$ is finite when $h \equiv 0$. This result may be improved as follows.

Proposition 5.1. *If* $\mathrm{Re} h(0)/g(0) < 1$, *then* $\sigma_{\mathrm{p}}(S_g)$ *is finite.*

Proof Let $\lambda \in \sigma_{\mathrm{p}}(S_g)$. Then, by Theorem 5.2, $\lambda = \dfrac{g(0)}{k}$, where $k \in \mathbb{N}$ satisfies

$$\mathrm{Re}\left(\frac{a_j}{g(0)}\right) < \frac{1}{kp} \qquad \text{for } j = 1, \ldots, m.$$

Summation yields

$$\mathrm{Re}\left(\frac{g(0) - h(0)}{g(0)}\right) = \mathrm{Re}\left(\frac{a_1 + \cdots + a_m}{g(0)}\right) < \frac{m}{kp}$$

and therefore

$$0 < 1 - \mathrm{Re}\frac{h(0)}{g(0)} < \frac{m}{kp}.$$

Thus

$$k < \frac{m}{p}\left(1 - \mathrm{Re}\frac{h(0)}{g(0)}\right)^{-1}.$$

This ensures that $\sigma_{\mathrm{p}}(S_g)$ is finite. □

In remarkable contrast to the case of the classical Cesàro operator, $\sigma_{\mathrm{p}}(S_g)$ may well be infinite. The preceding theorems lead, for instance, to the following result.

Example 5.1. If $g(z) := \dfrac{1+\mathrm{i}}{1-z} + \dfrac{1-\mathrm{i}}{1-\mathrm{i}z} - 3$, then $\sigma_{\mathrm{p}}(S_g) = \left\{-\dfrac{1}{k} : k \in \mathbb{N}\right\}$.

6. Subnormality and hyponormality

In this section, we consider the question of subnormality and hyponormality for certain classes of generalized Cesàro operators. Our first result describes the only class of such operators for which we know that subnormality holds.

Example 6.1. If $g(z) = \dfrac{1}{1-z}$, then S_g is the classical Cesàro operator C, and hence subnormal on $H^2(\mathbb{D})$. Moreover, if $g(z) = \dfrac{1}{1-bz}$ for some $b \in \mathbb{C}$

with $|b| = 1$, then S_g is unitarily equivalent to C and therefore subnormal on $H^2(\mathbb{D})$.

Proof As noted earlier, the first statement is a celebrated result due to T. L. Kriete and D. Trutt [14]. To establish the second claim, we consider, as before, $(U_b f)(z) := f(bz)$. Then U_b is a surjective isometry on $H^2(\mathbb{D})$ with $S_g = U_b C U_{\bar{b}}$. Alternatively, in terms of the representing matrices

$$[S_g] = \begin{bmatrix} 1 & 0 & 0 & 0 & \dots \\ \frac{b}{2} & \frac{1}{2} & 0 & 0 & \dots \\ \frac{b^2}{3} & \frac{b}{3} & \frac{1}{3} & 0 & \dots \\ \frac{b^3}{4} & \frac{b^2}{4} & \frac{b}{4} & \frac{1}{4} & \dots \\ \vdots & \vdots & \vdots & \vdots & \ddots \end{bmatrix} \quad \text{and} \quad [U_b] = \begin{bmatrix} 1 & 0 & 0 & 0 & \dots \\ 0 & b & 0 & 0 & \dots \\ 0 & 0 & b^2 & 0 & \dots \\ 0 & 0 & 0 & b^3 & \dots \\ \vdots & \vdots & \vdots & \vdots & \ddots \end{bmatrix},$$

we have the identities

$$[S_g]\,[U_b] = \begin{bmatrix} 1 & 0 & 0 & 0 & \dots \\ \frac{b}{2} & \frac{b}{2} & 0 & 0 & \dots \\ \frac{b^2}{3} & \frac{b^2}{3} & \frac{b^2}{3} & 0 & \dots \\ \frac{b^3}{4} & \frac{b^3}{4} & \frac{b^3}{4} & \frac{b^3}{4} & \dots \\ \vdots & \vdots & \vdots & \vdots & \ddots \end{bmatrix} = [U_b]\,[C]\,.$$

This establishes the desired unitary equivalence. □

We now turn to two classes of operators which are intimately related to the classical Cesàro operator. We shall see that hyponormality fails for the first class, but holds for the second one. This seems to indicate that it may be difficult to characterize the symbols for which the corresponding generalized Cesàro operator is hyponormal or even subnormal.

By contrast, as will be discussed in the next section, many types of generalized Cesàro operators belong to the class of subdecomposable operators. In fact, we are not aware of any generalized Cesàro operator that fails to be subdecomposable. Thus it seems reasonable to conjecture that all Cesàro operators with rational symbol are subdecomposable.

Example 6.2. If $g(z) = \dfrac{1}{1 - z^n}$ for some $n \geq 2$, then S_g is <u>not</u> hyponormal on $H^2(\mathbb{D})$.

Proof For an arbitrary polynomial p of degree n with distinct roots $\zeta_1, \ldots, \zeta_n \in \mathbb{C}$, we have the partial fraction decomposition

$$\frac{1}{p(z)} = \sum_{k=1}^{n} \frac{1}{p'(\zeta_k)(z - \zeta_k)}.$$

Consequently,

$$\frac{1}{1 - z^n} = \frac{1}{n} \sum_{k=1}^{n} \frac{1}{1 - \mathrm{e}^{-2k\pi \mathrm{i}/n}\, z}.$$

Thus our symbol g has the standard representation with

$$a_1 = \cdots = a_n = \frac{1}{n} \quad \text{and} \quad h \equiv 0.$$

It is not difficult to compute the matrix representation for $S_g^* S_g - S_g S_g^*$ and then to check, either by hand or by using a computer algebra system such as *Mathematica*, that this matrix fails to be positive definite for specific values of $n \geq 2$. However, the following systematic approach will settle the issue immediately for all $n \geq 2$. By the formula for $\sigma(S_g)$ from Theorem 5.3, the spectral radius of our operator S_g satisfies $r(S_g) \leq 1$. But

$$(S_g 1)(z) = \frac{1}{z} \int_0^z \frac{1}{1 - \omega^n}\, \mathrm{d}\omega = \sum_{k=0}^{\infty} \frac{1}{n\,k + 1}\, z^{n\,k},$$

and therefore

$$\|S_g 1\|^2 = \sum_{k=0}^{\infty} \frac{1}{(n\,k + 1)^2} > 1 = \|1\|^2,$$

which shows that $\|S_g\| > 1$. Since, for hyponormal operators, the norm and the spectral radius coincide, we infer that S_g cannot be hyponormal on $H^2(\mathbb{D})$. □

For the symbol $h(z) = (1+z)/(1-z)$, it was recently shown in [2] that S_h is hyponormal, which provided a partial solution to a problem posed by A. G. Siskakis [25]. In the next example, the approach from [2] will be extended to cover arbitrary convex combinations of the classical Cesàro operator and the operator considered by A. G. Siskakis. It would be interesting to know if these operators are actually subnormal. Note that, in general, a convex combination of two hyponormal operators need not be hyponormal, so the following result requires some work.

Example 6.3. If $g(z) = \dfrac{1 + t\,z}{1 - z}$ for arbitrary $0 \leq t \leq 1$, then S_g is hyponormal on $H^2(\mathbb{D})$.

Proof First observe that

$$g(z)=\frac{1}{1-z}+\frac{tz}{1-z}=\sum_{k=0}^{\infty}z^k+t\sum_{k=1}^{\infty}z^k=\sum_{k=0}^{\infty}c_k\,z^k$$

with $c_0=1$ and $c_k=1+t$ for $k\geq 1$. Thus

$$[S_g]=\begin{bmatrix} 1 & 0 & 0 & 0 & 0\ldots \\ \frac{1+t}{2} & \frac{1}{2} & 0 & 0 & 0\ldots \\ \frac{1+t}{3} & \frac{1+t}{3} & \frac{1}{3} & 0 & 0\ldots \\ \frac{1+t}{4} & \frac{1+t}{4} & \frac{1+t}{4} & \frac{1}{4} & 0\ldots \\ \frac{1+t}{5} & \frac{1+t}{5} & \frac{1+t}{5} & \frac{1+t}{5} & \frac{1}{5}\ldots \\ \vdots & \vdots & \vdots & \vdots & \vdots\ddots \end{bmatrix}.$$

Our task is to verify that the infinite matrix

$$M:=\left[S_g^*S_g-S_gS_g^*\right]=(m_{j,k})_{j,k\geq 0}$$

is positive semi-definite. For $0\leq j\leq k$, a routine computation leads to

$$m_{j,k}=m_{k,j}=(1+t)\left(\sum_{\nu=k+1}^{\infty}\frac{1+t}{\nu^2}-\frac{1+t}{k+1}-\frac{t}{(k+1)^2}+\frac{t}{(j+1)(k+1)}\right).$$

Also, note that

$$\sum_{\nu=1}^{\infty}\frac{1}{\nu^2}=\frac{\pi^2}{6},$$

but the matrix M seems to be difficult to handle, even with the help of *Mathematica*. To proceed further, we use an idea from [2], namely telescoping:

$$\frac{1}{k+1}=\sum_{\nu=k+1}^{\infty}\left(\frac{1}{\nu}-\frac{1}{\nu+1}\right)=\sum_{\nu=k+1}^{\infty}\frac{1}{\nu(\nu+1)}$$

$$\frac{1}{(k+1)^2}=\sum_{\nu=k+1}^{\infty}\left(\frac{1}{\nu^2}-\frac{1}{(\nu+1)^2}\right)$$

It follows that

$$m_{j,k}=m_{k,j}=(1+t)\left(\sum_{\nu=k+1}^{\infty}\frac{(1-t)\nu+1}{\nu^2(\nu+1)^2}+\frac{t}{(j+1)(k+1)}\right).$$

Since $0 \le t \le 1$, it remains to establish positive semi-definiteness for the infinite matrices

$$Q := \left(\frac{1}{(j+1)(k+1)} \right)_{j,k \ge 0} \qquad \text{and} \qquad R := (r_{j,k})_{j,k\ge 0}$$

with the entries

$$r_{j,k} = r_{k,j} = s_k := \sum_{\nu=k+1}^{\infty} \frac{(1-t)\nu + 1}{\nu^2(\nu+1)^2} \qquad \text{for } 0 \le j \le k.$$

For the matrix Q, the task is easy. Indeed, let

$$h(z) := \sum_{k=0}^{\infty} \frac{1}{k+1}\, z^k \qquad \text{for all } z \in \mathbb{D},$$

observe that $h \in H^2(\mathbb{D})$, and define q on $H^2(\mathbb{D})$ by

$$q(f) := \langle f, h \rangle\, h \qquad \text{for all } f \in H^2(\mathbb{D}).$$

Then $\langle q(f), f\rangle = |\langle f, h\rangle|^2 \ge 0$ for all $f \in H^2(\mathbb{D})$. Since, obviously, $[q] = Q$, it follows that Q is positive semi-definite. To settle the case of the matrix R, we compute the principal minors of R as follows:

$$\det R_n = \begin{vmatrix} s_0 & s_1 & s_2 & \dots & s_n \\ s_1 & s_1 & s_2 & \dots & s_n \\ s_2 & s_2 & s_2 & \dots & s_n \\ \vdots & \vdots & \vdots & \ddots & \vdots \\ s_n & s_n & s_n & \dots & s_n \end{vmatrix} = \begin{vmatrix} s_0 - s_1 & 0 & 0 & \dots & 0 \\ s_1 & s_1 & s_2 & \dots & s_n \\ s_2 & s_2 & s_2 & \dots & s_n \\ \vdots & \vdots & \vdots & \ddots & \vdots \\ s_n & s_n & s_n & \dots & s_n \end{vmatrix}$$

$$= \begin{vmatrix} s_0 - s_1 & 0 & 0 & \dots & 0 \\ s_1 - s_2 & s_1 - s_2 & 0 & \dots & 0 \\ s_2 & s_2 & s_2 & \dots & s_n \\ \vdots & \vdots & \vdots & \ddots & \vdots \\ s_n & s_n & s_n & \dots & s_n \end{vmatrix} = \dots = \begin{vmatrix} s_0 - s_1 & 0 & 0 & \dots & 0 \\ s_1 - s_2 & s_1 - s_2 & 0 & \dots & 0 \\ s_2 - s_3 & s_2 - s_3 & s_2 - s_3 & \dots & 0 \\ \vdots & \vdots & \vdots & \ddots & \vdots \\ s_n & s_n & s_n & \dots & s_n \end{vmatrix}$$

Consequently, we have

$$\det R_n = s_n \prod_{k=1}^{n} \frac{(1-t)k + 1}{k^2(k+1)^2} \ge 0$$

for all $n \in \mathbb{N}$, which shows that R is indeed positive semi-definite. □

7. Subdecomposability

A remarkable accomplishment due to E. Albrecht and J. Eschmeier [1] from 1997 is the following intrinsic characterization of subdecomposable operators. Here $T \in \mathcal{L}(X)$ denotes an arbitrary bounded linear operator on a complex Banach space X.

Theorem 7.1. *T is subdecomposable precisely when T satisfies Bishop's property (β), in the sense that, for each open set $U \subseteq \mathbb{C}$ and each sequence of analytic functions $f_n : U \to X$ for which*

$$(T - \lambda I) f_n(\lambda) \to 0 \quad \text{as} \quad n \to \infty,$$

locally uniformly on U, it follows that

$$f_n(\lambda) \to 0 \quad \text{as} \quad n \to \infty,$$

again locally uniformly on U. Moreover, T is decomposable if and only if both T and T^* have Bishop's property (β).

To verify Bishop's property (β) in concrete cases, the following result often turns out to be useful; see Theorem 1.7.4 of [15]. Basically, the result says that suitable control on the growth of one-sided resolvents for an operator entails property (β) and hence subdecomposability.

Theorem 7.2. *Suppose that $\sigma(T)$ is a closed disc, let V be an open neighborhood of $\sigma(T)$, and suppose that there exist a finite (or at least totally disconnected) set $E \subseteq \partial\sigma(T)$, a locally bounded function $\omega : V \setminus E \to (0, \infty)$, and an increasing function $\gamma : (0, \infty) \to (0, \infty)$ such that*

(a) $\gamma(\operatorname{dist}(\lambda, \partial\sigma(T)) \,\|x\| \leq \omega(\lambda)\,\|(T - \lambda I)x\|$ *for all $x \in X$ and $\lambda \in V \setminus \partial\sigma(T)$;*

(b) $\int_0^c \log|\log \gamma(t)|\, dt < \infty$ *for some $c > 0$.*

Then T is subdecomposable.

The following theorem summarizes results from [18] and [20]. The proof of the first assertion is based on Theorem 7.2, but the other cases require different methods. Some of the perturbation techniques that are relevant in this context are due to E. Albrecht and W. J. Ricker [3] after earlier work of L. Zsidó [30]. It remains an interesting open problem if all generalized Cesàro operators with rational symbol are subdecomposable.

Theorem 7.3. *In each of the following cases, S_g is subdecomposable on $H^p(\mathbb{D})$ for arbitrary $1 < p < \infty$:*

$\star \quad g(z) = \dfrac{a}{1 - b\,z} \quad$ *for* $a, b \in \mathbb{C}$ *with* $|b| = 1$

$\star \quad g(z) = \dfrac{1+z}{1-z} = \dfrac{2}{1-z} - 1$

$\star \quad g(z) = \dfrac{1}{1-z^2} = \dfrac{1/2}{1-z} + \dfrac{1/2}{1+z}$

$\star \quad g(z) = \dfrac{z^m}{1-z} \quad$ *for arbitrary* $m \in \mathbb{N}$

$\star \quad g(z) = \dfrac{\varphi(z)}{1-z} \quad$ *where* φ *is analytic on* $\overline{\mathbb{D}}$ *with* $\mathrm{Re}\varphi > 0$ *on* $\mathbb{D}$

8. The case of weighted Bergman spaces

We conclude with a brief discussion of generalized Cesàro operators on the classical Bergman space and its weighted analogues. For $-1 < \alpha < \infty$ and $1 \leq p < \infty$, we consider the *weighted Bergman space* $L^{p,\alpha}_a(\mathbb{D})$ consisting of all $f \in H(\mathbb{D})$ for which

$$\int_{\mathbb{D}} |f(z)|^p \left(1 - |z|^2\right)^\alpha d\mu(z) < \infty,$$

where μ denotes planar Lebesgue measure on $\mathbb{D}$. For an attractive account of the theory of Bergman spaces, we refer to [13]. The classical Bergman space corresponds, of course, to the choice $\alpha = 0$ and $p = 2$, while, for fixed $1 \leq p < \infty$, the Hardy space $H^p(\mathbb{D})$ may be viewed as the limiting case of the spaces $L^{p,\alpha}_a(\mathbb{D})$ as $\alpha \to -1^+$; see, for instance, [2].

The state of the art for Cesàro-type operators in the Bergman setting may be summarized as follows:

- The theory of the spectral picture extends to the case of generalized Cesàro operators on weighted Bergman spaces when $p > 1$; see [2], [6], [8], [19], and [29].

- The classical Cesàro operator C fails to be hyponormal on the classical Bergman space $L^{2,0}_a(\mathbb{D})$, but C is subdecomposable on $L^{p,\alpha}_a(\mathbb{D})$ for all $\alpha > -1$ and $p > 1$; see [16] and [17].

- Not much is known about subdecomposability of generalized Cesàro operators in this setting, and the case $p = 1$ is wide open.

References

1. E. Albrecht and J. Eschmeier, Analytic functional models and local spectral theory, *Proc. London Math. Soc. (3)* **75** (1997), 323–348.

2. E. Albrecht, T. L. Miller, and M. M. Neumann, Spectral properties of generalized Cesàro operators on Hardy and weighted Bergman spaces, *Arch. Math. (Basel)* ***85*** (2005), 446–459.
3. E. Albrecht and W. J. Ricker, Local spectral theory for operators with thin spectrum, in *Spectral Analysis and its Applications,* Theta Ser. Adv. Math., Theta Foundation, Bucharest, Romania, 2003, pp. 1–25.
4. A. Aleman and J. A. Cima, An integral operator on H^p and Hardy's inequality, *Journal d'Analyse Mathématique* **85** (2001), 157–176.
5. A. Aleman and A. G. Siskakis, An integral operator on H^p, *Complex Variables Theory Appl.* **28** (1995), 149–158.
6. A. Aleman and A. G. Siskakis, Integration operators on Bergman spaces, *Indiana Univ. Math. J.* **46** (1997), 337–356.
7. A. Brown, P. R. Halmos, and A. L. Shields, Cesàro operators, *Acta Sci. Math. (Szeged)* **26** (1965), 125–137.
8. A. Dahlner, Decomposable extension of the Cesàro operator on the weighted Bergman space and Bishop's property (β), in *Some Resolvent Estimates in Harmonic Analysis,* Dissertation, Lund University, Lund, Sweden, 2003, pp. 73–100.
9. P. L. Duren, *Theory of H^p Spaces,* Academic Press, New York, 1970.
10. M. González, The fine spectrum of the Cesàro operator in ℓ^p $(1 < p < \infty)$, *Arch. Math. (Basel)* **44** (1983), 355–358.
11. G. H. Hardy, Note on a theorem of Hilbert, *Math. Z.* **6** (1920), 314–317.
12. G. H. Hardy, J. E. Littlewood, and G. Pólya, *Inequalities,* Cambridge University Press, Cambridge, 1934.
13. H. Hedenmalm, B. Korenblum, and K. Zhu, *Theory of Bergman Spaces,* Graduate Texts in Math. **199**, Springer-Verlag, New York, 2000.
14. T. L. Kriete and D. Trutt, The Cesàro operator on ℓ^2 is subnormal, *Amer. J. Math.* **93** (1971), 215–225.
15. K. B. Laursen and M. M. Neumann, *An Introduction to Local Spectral Theory,* Clarendon Press, Oxford, 2000.
16. T. L. Miller and V. G. Miller, On the approximate point spectrum of the Bergman space Cesàro operator, *Houston J. Math.* **27** (2001), 479–494.
17. T. L. Miller and V. G. Miller, The Cesàro operator on the Bergman space $A^2(\mathbb{D})$, *Arch. Math. (Basel)* **78** (2002), 409–416.
18. T. L. Miller, V. G. Miller, and M. M. Neumann, Growth conditions, compact perturbations, and operator subdecomposability, with applications to generalized Cesàro operators, *J. Math. Anal. Appl.* **301** (2005), 32–51.
19. T. L. Miller, V. G. Miller, and M. M. Neumann, Analytic bounded point evaluations for rationally cyclic operators on Banach spaces, *Integr. Equ. Oper. Theory* **51** (2005), 257–274.
20. T. L. Miller, V. G. Miller, and R. C. Smith, Bishop's property (β) and the Cesàro operator, *J. London Math. Soc. (2)* **58** (1998), 197–207.
21. Ch. Pommerenke, Schlichte Funktionen und analytische Funktionen von beschränkter mittlerer Oszillation, *Comment. Math. Helv.* **52** (1977), 591–602.
22. M. Putinar, Hyponormal operators are subscalar, *J. Operator Theory* **12**

(1984), 385–395.

23. A. G. Siskakis, Composition semigroups and the Cesàro operator on H^p, *J. London Math. Soc. (2)* **36** (1987), 153–164.
24. A. G. Siskakis, The Cesàro operator is bounded on H^1, *Proc. Amer. Math. Soc.* **110** (1990), 461–462.
25. A. G. Siskakis, The Koebe semigroup and a class of averaging operators on $H^p(\mathbb{D})$, *Trans. Amer. Math. Soc.* **339** (1993), 337–350.
26. A. G. Siskakis, On the Bergman space norm of the Cesàro operator, *Arch. Math. (Basel)* **67** (1996), 312–318.
27. S. W. Young, *Algebraic and Spectral Properties of Generalized Cesàro Operators,* Dissertation, University of North Carolina, Chapel Hill, 2002.
28. S. W. Young, Spectral properties of generalized Cesàro operators, *Integr. Equ. Oper. Theory* **50** (2004), 129–146.
29. S. W. Young, Generalized Cesàro operators and the Bergman space, *J. Operator Theory* **52** (2004), 341–351.
30. L. Zsidó, Invariant subspaces of linear operators on Banach spaces, *J. Operator Theory* **1** (1979), 225–260.

Michael M. Neumann
Department of Mathematics and Statistics
Mississippi State University
Mississippi State, MS 39762, USA

neumann@math.msstate.edu

TRIBUTE TO MIGUEL DE GUZMÁN: REFLECTIONS ON MATHEMATICAL EDUCATION CENTERED ON THE MATHEMATICAL ANALYSIS

BALDOMERO RUBIO SEGOVIA
Department of Mathematical Analysis
University Complutense
Madrid, Spain
E-mail: brubio@mat.ucm.es

Miguel had accepted to take part in this Conference, and there is no doubt that we all would have had the opportunity to learn and enjoy his words and teaching. But he passed away last April, and I have been asked the impossible task of replacing him to some extent. I have interpreted that it was to remember him and carry out an analysis, of necessity short, of his huge work. I think that the only reason for which I have been appointed to be present in this event is that I had the privilege of spending thirty-five years working closely with him in many periods of our lives which ended up in several books and research works.

Miguel started teaching at the Universidad Complutense de Madrid in September 1969. He had just finished his PhD in Chicago with Alberto Calderón and came back to Spain full of ideas and hoping to recreate the stimulating and creative university atmosphere that he had experienced in Chicago. Although mentioned on many occasions, it is important to repeat once again that we used to carry out an inert study which did not favor the courage to do research or to think in solving attainable problems. Miguel introduced Mathematics as a live object where it was possible to take part and discover new things. Many theses were soon supervised by him, being mine the first of all of them, and for which I am very grateful. Shortly before, we had just published an article in American Mathematical Monthly, which showed, somehow, the gradually increasing desire that was emerging in his environment and which made his work internationally known at the end of his life.

Recently, I had the honor to participate in a ceremony organized by the Universidad Internacional Menéndez Pelayo as a tribute to him in Santander. At the end of the ceremony, the rector presented to Mayte, his wife since 1971, the Medal of Honor that the senate had agreed to award Miguel with. In this ceremony in Santander, several people have glossed various aspects of his life and work developed in different periods. We can also find some other mentions, mine included, in the issue of July of Suma journal and something similar will be published in the next issue of the Gaceta de la Real Sociedad Española. It goes without saying that there are many people doing their best to honor our friend Miguel de Guzmán. The Faculty of Mathematics of the Universidad Complutense will pay tribute to him next December. There is a widespread wish of making his personality and work widely recognized so the future generations will also benefit from them.

At this point, I would like to emphasize two of his books that made him being considered one of the leading mathematicians all over the world: the monograph known as the "yellow book" entitled

- *Differentiation of Integrals in* $\mathbb{R}^n$ (in Lecture Notes in Mathematics, Springer-Verlag, 1975)

and the monograph known as the "blue book" entitled

- *Real variable methods in Fourier Analysis* (in North-Holland Mathematics Studies, 1981).

These two books are enough to consider Miguel as one of the most influential and international known Spanish mathematicians in the 20th century. For those interested in obtaining more information about the meaning of these books and, in general, about Miguel himself as a researcher, I recommend the article written by Fernando Soria, professor of the Universidad Autónoma de Madrid, published in the above mentioned issue of the Gaceta de la Real Sociedad Matemática Española.

However, Miguel did not limit his production to the Differentiation of Integrals and the Fourier Analysis. He constantly worked in the study of new fields, for their diffusion, to contribute with new ideas along with some co-workers. A good example of this, is the book

- *Fractal structures and applications*, published by Labor Matemáticas in 1993.

He wrote it in collaboration with Miguel Ángel Martín, Manuel Morán and Miguel Reyes. He had lately focused his interest on Tensegrity, a very recent discipline affecting Architecture and Art. He had almost finished a monograph on this topic at the moment of his death.

In addition to the textbooks for secondary school students written in collaboration with José Colera and Adela Salvador, Miguel also wrote the following texts for undergraduate students:

- *Ecuaciones diferenciales ordinarias. Teoría de estabilidad y control.* (Ordinary differential equations: stability and control theory). Alhambra, 1975.
- *Integración: teoría y técnicas.* (Integration: theory and techniques). Alhambra, 1979.
- *Problemas, conceptos y métodos del Análisis Matemático.* (Problems, concepts and methods of mathematical analysis). Pirámide, 1990-1993.

And more recently:

- *Cómo hablar, demostrar y resolver en Matemáticas.* (How to speak, prove and solve in Mathematics). Anaya, 2004.

His books are worthy of admiration due to his intention of transmitting to a wide public his vision of Mathematics, and particularly of geometry, a field which had been a lifelong interest for him. The main objective was to show the educational value of this science, its enjoyable aspects and the usefulness of the effort made to deal with the stimulating problems that may be set out. The titles in this section are as follows:

- *Mirar y ver.* (Look and see). Alhambra, 1977.
- *Cuentos con cuentas.* (The Countingbury Tales). Labor, 1984.
- *Aventuras matemáticas.* (Mathematical adventures). Labor, 1986.
- *Para pensar mejor.* (To think better). Labor, 1991.
- *Aventuras matemáticas: una ventana hacia el caos y otros episodios.* (Mathematical Adventures: A window opened to chaos and other topics). Pirámide, 1995.
- *El rincón de la pizarra.* (The blackboard corner). Pirámide, 1996.
- *La experiencia de descubrir en geometría.* (The experience of discovering in geometry). Nivola, 2002.
- *Mirar y ver.* (Look and see). Nivola, 2004.

And it is also remarkable his book

- *Los Espingorcios*, Labor, 1984,

an imaginative and delicious tale written for his children, Miguel and Mayte. His creativity went beyond buying tales for his children, he made them up. Moreover, Miguel wrote

- *Los matemáticos no son gente seria.* (Mathematicians are not serious people). Rubes, 1996,

in collaboration with Claudi Alsina, in which he highlights his love for the entertaining and enjoyable aspects of math.

His concern with the mathematical education, and especially in that regarding secondary school, took a long period of his life. He served as President in the International Commission for Mathematical Instruction (ICMI) from 1991 to 1998, and in July 1996, he took part in the 8th International Congress on Mathematical Education organized by this Commission in Seville. Nevertheless, his continuous teaching in Teachers' Centers of any place in Spain is what better describes his willingness to encourage and supervise the work of many teachers, not only in Spain but also in several Latin American countries. He did not relax at weekends, he often used to travel in his own car far away from Madrid to live for several hours with other teachers. His web page has been proudly visited by many people all over the word and a big part of its content was collected in a CD entitled

- *Pensamientos en torno al quehacer matemático* (Thoughts about mathematical tasks).

Miguel had edited this CD himself and used to offer it for free to whom agreed to make a donation to any NGO interested in educational topics. He never asked for the proof of this commitment and did not want to know how much they donated. We are still receiving requests of the CD from people visiting the web page.

I do not want to finish this memory of Miguel de Guzmán without mentioning his unique personality. His main motivation was to devote himself to everybody else through the Mathematics. First of all, by giving guidance to those young people that he met at his arrival in Madrid in 1969 when they wanted to begin their university studies, or by promoting the PhD in North American Universities for recently graduate students that when coming back to Spain, have crucially contributed to transform the way of working with Mathematics. Secondly, by brilliantly teaching in many different areas and being always ready to help whoever needed some advice or encouragement. Thirdly, by getting the media to detect and strengthen the mathematical talent of the youth with the programs sponsored by Vodafone in the Spanish Royal Academy of Mathematical, Physical and Natural Sciences. And finally, by creating and encouraging a NGO -named CUES- to help in the field of Mathematics in third-world countries, and many other activities. Because of his constant dedication to others and because of his calm, cheerful and dynamic character, his memory and work will always be present among us.

Now, the duty of all of us who have received this valuable heritage consists of extending it as much as possible so it can be enjoyed by those who did not have the opportunity to meet this fascinating man personally.

∼ · ∼ · ∼ · ∼ · ∼ · ∼ · ∼ · ∼ · ∼ · ∼

My contribution to this Conference, from the point of view of the mathematical analysis, consists of some reflections on mathematical education that I personally think they are very close to Miguel's and which deal with pre-university teaching, a topic in which I have been interested in several periods of my professional life.

In 1961, I finished my degree in Mathematics and was able to carry out my dedication to help freshmen and fulfil my duties as a grant holder in high schools teaching training along with the excellent professor José Ramón Pascual Ibarra, those days teacher in the Instituto de Bachillerato Cervantes of Madrid. Pedro Puig Adam had also been a middle-education teacher in the same period and I was lucky enough to meet him, just in the same year he died, when I was an undergraduate student and, for that reason, I had the opportunity to attend some of the classes he taught to students of Bachillerato in the Instituto de San Isidro. The students were ten years old and it was a once in a lifetime experience for me to see the teacher working with them, as it was a long talk we had in his house in Madrid along with two other classmates of the faculty.

The academic year 61/62 begun with a pilot program in Mathematics addressed to students of Bachillerato from Granada, Barcelona and Madrid. The program in Madrid was supervised by Pascual Ibarra, my tutor of the above mentioned grant in teaching training. We were three grant holders and we used to attend the professor's classes and took part occasionally in some of them. His teaching method was excellent to learn the skills necessary for teachers even more when he had such a wide ability and a vast experience on the subject. This fact contributed positively to my teacher training.

Those were the times of modern mathematical revolution, considered nowadays of doubtful usefulness for children's mathematical learning.

That pilot program was indeed the first attempt to put into practice this model. The stress was placed in the structures and in the set theory as well as in equivalence and order relations. Integer and rational numbers were discovered and presented as collections of plan dots in parallels or concurrent lines, depending on the case. The multiplication of integer numbers was shown as an arithmetic dogma. The aim was that the objects which responded to the same structure emerged from it, as if it would not be true that we are able to understand what a structure is after doing specific constructions which respond to it and we personally make the correct abstraction. The process of abstraction is something that nobody should do for someone else. It is also true that the pilot program contained a combinatorial analysis that stood out the great figure of Pascual Ibarra because the real problems and the guidelines for the solution were then set.

The modernity of mathematical teaching spread from that pilot experience. But Puig Adam did not have the chance to see the development of this pretending reform. However, he managed to express his concern about the new tendencies as we can see in the preface of the book, published in October 1959, addressed to the students enrolled in the pre-university year:

"In the area of sciences, the new regulations of the pre-university year put an end to one of the boldest and finest pedagogical essays ever promoted by the Spanish Ministry of Education. The aim of this paper is not to analyze or regret the different causes that justify this change but rather to assimilate its reasons and to try that the students can benefit as much as possible from this new direction. Under the undoubted pressure of Spain's technological needs, the aim is to achieve a course with a stressed informative content without reducing the formative values that encouraged its previous organization, still remaining in the area of humanities".

As it can be appreciated, Puig Adam observes with concern this new regulation for the pre-university year. However, he tries to keep to it for the benefit of the students.

In the mid 1960s, I had the opportunity to teach Bachillerato students for two years. Teachers had to propose some text books and I chose undoubtedly those by Rey Pastor and Puig Adam, which do not deal with modern Mathematics at all. I rescued these books from my library and I eagerly read them again. They belong to the 1957's syllabus, slightly before the advent of the modernization of Mathematics. I remember the times when I taught 10-year-old girls by following these books. For instance, when it came to calculating square roots, the books suggested getting some square cards that could be split up into ten identical strips and, if necessary, these strips would be divided into smaller squares. We created the biggest possible square with a certain material, so the algorithm resulting from the calculation done could be deduced. These girls thus learnt how to calculate square roots with and without cards.

Another example taken from these books consisted in multiplying two integer numbers by thinking of a train travelling from Barcelona to Madrid, or vice versa, and crossing Calatayud (a smaller city between the two above mentioned). Calatayud was the origin whereas Madrid was in the negative side and Barcelona in the positive one. The train was in Calatayud at the starting point. Speed, time and position had their own + or -sign, and the rest was easy to figure out. In order to add something new about integer numbers and their arithmetic, my personal point of view on this matter consists in not just stressing its construction as a quotient set but rather in studying the $a-b$ differences in natural numbers, interpreting them as an x number where $b+x$ is a, in the case that a is greater than b. If the sum $d+y$ is c, it is capital to obtain the sum $x+y$ and the product xy in terms of a, b, c and d, and remark that the sum $x+y$ and the product xy are also differences expressed by a, b, c and d.

In the academic year 67/68, I stopped teaching in high schools and devoted myself to teaching at universities. Nevertheless, I was interested -and in fact, I have always been- in Secondary Education, and consequently I wrote the book

- *Iniciación a la matemática superior* (Introduction to Advanced Mathematics), published by Alhambra in 1969.

This book aimed to serve as a bridge between the secondary and the university education. I should mention that this book was a complete success and received very good critics. I am especially glad that I put on it enough notions of astronomy to get to know the sky-blue sphere, the movement of the stars, the Earth's rotation and translation movements and the measure of time. This book was aimed at the last year of secondary education. However, it is not compulsory to learn this type of general knowledge in any level of education nowadays.

I strongly believe that Spanish students face a critical situation in their mathematical education, from elementary to university levels. This crisis can worsen in the future if we do not try to solve it immediately. I will now present some of my thoughts and provide significant data in order to describe the symptoms where this crisis becomes apparent.

As far as elementary education is concerned, there has never existed such a deficit regarding the teachers' training. We have shifted from a 7-year to a 2-year Bachillerato in only a couple of decades. In the past, Puig Adam could teach an 11-year student but today, the formal requirements for a Mathematics teacher in elementary education are very low. Besides, there are some sociological factors -both in elementary and secondary education- which are detrimental to education, such as a TV excess of information as well as in other audiovisual media. This makes students find it difficult to concentrate on a problem and develop a creative mental activity of their own. On the other hand, teaching Mathematics is compulsory in schools and this leads to the hostility of children towards learning. The results are general school failure, Spanish students scoring very low in the international ranking and wanting less and less to study a university degree in Mathematics.

Allow me to explain an anecdote concerning this last aspect. Last August, I was invited by the Universidad Internacional Menéndez Pelayo to give a lecture in front of fifty students who were planning to start their university education. They all were nationwide selected because of their excellent academic records. During 15 days, they were offered a wide intellectual perspective in the Palacio de la Magdalena, Santander. I was supposed to offer an outlook on the meaning of Mathematics and I did it by offering some particular examples of how Mathematics has contributed to the new inventions along the History. Once my lecture was over, we had a long discussion that let me find out the deficits they have had in their recently-finished secondary education. None of the students was planning to study a university degree in Mathematics but someone said that they would have considered studying Mathematics, had they been aware of what Mathematics can provide in many aspects.

I will now try to be as accurate as possible regarding my interpretation of the deficits in secondary education and that is why I will give my point of view from the area I know best: the Mathematical Analysis. I think that the mental activity that mathematical learning implies consists essentially in facing problematic situations where, first of all, it is necessary to clearly understand what it is all about in order to set mechanisms of research which will permit to find a solution. What is very serious is not the fact of not being able to find a solution but rather not being able to initiate a research activity. This usually happens to the students who believe that they have to learn by heart a list of formulae. In other words, these students think that someone will teach them how to solve a problem so their only duty consists in trying to solve other similar problem later on.

The success of this program highly depends on the contents the teacher is obliged to teach. Syllabuses play a key role but, how are syllabuses chosen? I have realized that mathematical syllabuses are sometimes chosen according to a personal criterion instead of arising from a joint reflection in a committee and with the background knowledge of those who have the capacity to take part on it, such as the communities of educators and well-known experts in this field. It may be due to this initial fault, although it might have taken place long time ago, that the mathematical syllabuses in secondary education -and especially in Bachillerato- undergo serious faults. I will try to explain it within the area of Analysis, where I am more experienced in.

I consider that the educational value of Mathematics in secondary education lies in its capacity to develop the intelectual faculties of a person: abstraction, implication and logical equivalence, critics, accuracy when analyzing situations, agility when making decisions, error detection, among others. It is also almost certainly a necessary instrument for some human activities, but this effect may be a result of its formative value. Moving to a more specific topic, it is true that the theory of functions of a real variable is a basic tool in the study of other sciences, including social sciences, engineering or architecture. This does not necessarily imply that this matter has to be reflected in curricula of Bachillerato as it is nowadays. I justify this on the grounds of my belief that such matter has not an appropriate informative value at this educational stage. A student of Bachillerato is not still ready to study in depth complex processes such as integration, continuity o derivatives. The aim is to try to make them retain some formulae in order to do daily exercises instead of true problems. I have systematically confirmed this fact every time I try that my undergraduates immerse themselves in the secrets of infinitesimal calculation and face true problematic situations. Miguel de Guzmán, with whom I frequently used to talk about this matter, was on the same line. However, he considered that writing good books collaborating with Adela Salvador and José Colera was a manner of mitigating the bad effects of the syllabus they had to focus on.

At this stage, we can wonder: what is the best syllabus? We have some responsibility in this matter and therefore, we should wonder this question.

It would be advisable that teacher associations in secondary or university education -as well as any group of people interested in the matter- could analyze and make their own suggestions about this.

Here we can find some remarks and suggestions: we can intend to include integration and derivation elements in mathematical curricula of Bachillerato as they are necessary for physics. According to this view, we should value whether these subjects have little formative value -as it is commonly thought- or not within the context of Mathematics in Bachillerato. Another possibility is to include these elements in the physics curricula so that the students can understand what is meant by speed, acceleration or work. However, it is very usual that the contents learnt in Mathematics do not fit exactly in the contents learnt in physics lessons so students cannot link both things. I perfectly remember when I was in Bachillerato; it was in the physics lessons where I learnt the concepts of integrals and derivatives.

So, what could the mathematical analysis of Bachillerato mean? Let's think about what History has to say: infinitesimal calculation is a very elaborated product and its fundament comes after a long period where the mathematical problems set out did basically address simple infinite countable processes but were usually very difficult. As a first example, we can suggest what would happen when a container is filled up to its half in a first stage, and in the following stages it is filled half of what is left to be filled up completely. Has the container been filled up after any stage? What part has been filled up after two, three, four stages? This situation naturally leads to the series concept. We can also suggest the Fibonacci's rabbit problem, or even Jacob Bernoulli or Euler's problems dealing with the sum of the first n natural numbers around mid 18th century. Is there any sum able to express such sum? There is also the well-known Basel problem (Basel is Euler's hometown): What is the sum of the reciprocals of the squares of the positive integers? Many leading mathematicians of the day attacked the problem unsuccessfully but it was Euler who surprisingly solved it. Euler announced his discovering as follows:

"However, I have found now and against all odds a smart expression (for the sum of the reciprocals of the squares of the natural numbers) that depends on the quadrature of a circle. I have found out that six times this sum equals two times the length of circumference of a circle with diameter 1".

Would this type of countable finite or infinite processes be an appropriate subject-matter for Mathematics in Secondary School?

ON CERTAIN SPACES OF HOLOMORPHIC FUNCTIONS

MANUEL VALDIVIA *†

Departamento de Análisis Matemático
Universidad de Valencia
Dr. Moliner, 50
46100 Burjasot (Valencia)
Spain

Let Ω be a domain of holomorphy in $\mathbb{C}^n$. Let $\mathcal{H}(\Omega)$ denote the linear space over the field of complex numbers $\mathbb{C}$ formed by the holomorphic functions in Ω, endowed with the compact-open topology. In this paper, we construct an ordered nearly-Baire dense subspace of $\mathcal{H}(\Omega)$ whose non-zero elements cannot be extended holomorphically outside Ω. Some other subspaces of $\mathcal{H}(\Omega)$ with this same property are also constructed.

1. Introduction and notation

By $\mathbb{N}$ we represent the set of the positive integers and $\mathbb{N}_0 := \mathbb{N} \cup \{0\}$. By ω we denote the linear space $\mathbb{C}^{\mathbb{N}}$ of the sequences of complex numbers provided with the topology of convergence in each of the coordinates. We have that ω is a Fréchet space whose topological dual φ may be identified with those sequences (b_j) in $\mathbb{C}^{\mathbb{N}}$ whose terms are all zero from a certain position on, with the duality $\langle \omega, \varphi \rangle$ given by:

$$\langle (a_j), (b_j) \rangle := \sum_{j=1}^{\infty} a_j b_j, \quad (a_j) \in \omega, \quad (b_j) \in \varphi.$$

For a given $n \in \mathbb{N}$, let Ω be a domain in $\mathbb{C}^n$. Then $\mathcal{H}(\Omega)$ will be the linear space over $\mathbb{C}$ of the holomorphic functions in Ω, with the compact-open topology. If K is a compact subset of Ω, its convex holomorphic hull $\hat{K}$ is defined as

$$\{z \in \Omega : |f(z)| \leq \sup\{|f(u)| : u \in K\}, \forall f \in \mathcal{H}(\Omega)\}.$$

*Supported in part by MTM2005-08210

†2000 Mathematics Subject Classification: Primary 46F05, Secondary 46E10

Ω is said to be holomorphically convex whenever, for each compact subset K of Ω, $\hat{K}$ is compact.

In $\mathbb{C}^n$, the polydisk centered at $a = (a^{(1)}, a^{(2)}, ..., a^{(n)})$ with multiradius $r = (r^{(1)}, r^{(2)}, ..., r^{(n)})$, $r^{(j)} > 0$, $j = 1, 2, ..., n$, which we denote by $H(a, r)$, is defined as

$$\{z = (z^{(1)}, z^{(2)}, ..., z^{(n)}) \in \mathbb{C}^n : \mid z^{(j)} - a^{(j)} \mid < r^{(j)}, j = 1, 2..., n\}.$$

If r is a positive number, we write $H(a, r)$ for the polydisk with center in a and multiradius $(r, r, \overset{(n)}{\ldots}, r)$. If $z = (z^{(1)}, z^{(2)}, ..., z^{(n)}) \in \mathbb{C}^n$, we put

$$\mid z \mid := \sup\{\mid z^{(j)} \mid : \ j = 1, 2, ..., n\}.$$

We then have that $\mid \cdot \mid$ is a norm in $\mathbb{C}^n$. In what follows we shall assume that $\mathbb{C}^n$ is provided with this norm. It then happens that the open ball $B(a, r)$ with center a and radius r coincides with the polydisk $H(a, r)$. If $\alpha = (\alpha_1, \alpha_2, ..., \alpha_n) \in \mathbb{N}_0^n$ is a multiindex, its length will be represented by $\mid \alpha \mid = \alpha_1 + \alpha_2 + ... + \alpha_n$, also $\alpha! := \alpha_1!\alpha_2!...\alpha_n!$. Besides, if $z = (z^{(1)}, z^{(2)}, ..., z^{(n)})$, then $z^\alpha := (z^{(1)})^{\alpha_1}(z^{(2)})^{\alpha_2}...(z^{(n)})^{\alpha_n}$, with the agreement that, if $z^{(j)} = 0$ and $\alpha_j = 0$, then $(z^{(j)})^{\alpha_j} = 1$.

Let f be an element of $\mathcal{H}(\Omega)$ and let z_0 be a point in the boundary $\partial\Omega$ of Ω. We say that f is extendible holomorphically in z_0 if there exists $r > 0$ and a holomorphic function g in $B(z_0, r)$ which coincides with f in a connected component of $\Omega \cap B(z_0, r)$. If f is not holomorphically extendible in z_0, it is then said that z_0 is a completely singular point of f. When the set of all completely singular points of f coincides with $\partial\Omega$, we say that f does not extend holomorphically outside Ω.

If a domain Ω has the property that there is $f \in \mathcal{H}(\Omega)$ which does not extend holomorphically outside Ω, then Ω is said to be a domain of holomorphy. The theorem of Cartan-Thullen asserts the following: *A domain Ω is a domain of holomorphy if and only if it is holomorphically convex.*

It is well known that if Ω is a domain of holomorphy, the set M of all functions f in $\mathcal{H}(\Omega)$ which do not extend holomorphically outside Ω form a subset of the second category in the Fréchet space $\mathcal{H}(\Omega)$. In this paper, we are interested in constructing some subspaces of $\mathcal{H}(\Omega)$ such that, except the origin, they are contained in M. We obtain, for instance, ordered nearly-Baire spaces L, dense in $\mathcal{H}(\Omega)$, such that if $f \in L$, $f \neq 0$, f does not extend holomorphically outside Ω. Some other subspaces of $\mathcal{H}(\Omega)$ are also constructed satisfying this property.

Let (z_j) be a sequence in Ω. We shall say that this sequence has the interpolation property, whenever given an arbitrary sequence (a_j) of complex numbers, there is f in $\mathcal{H}(\Omega)$ such that $f(z_j) = a_j$, $j \in \mathbb{N}$. The following result is a particular case of a theorem found in [5] : *a) A sequence* (z_j) *in* Ω *of distinct elements has the interpolation property if and only if, for a given arbitrary compact subset* K *of* Ω, *there is* $n_0 \in \mathbb{N}$ *such that* $z_n \notin \hat{K}$, *for* $n \geq n_0$.

If (m,p) and (l,q) are in $\mathbb{N} \times \mathbb{N}$, we put $(m,p) < (l,q)$ if $m+p < l+q$, or, $m+p = l+q$ and $m < l$. Then $\leq$ is an order relation in $\mathbb{N} \times \mathbb{N}$. Given a double sequence $(z_{m,j})$ in Ω, we say that it has the interpolation property if, for an arbitrary double sequence $(a_{m,j})$ of complex numbers, there is $f \in \mathcal{H}(\Omega)$ such that $f(z_{m,j}) = a_{m,j}$, $m, j \in \mathbb{N}$. Clearly, $(z_{m,j})$ has the interpolation property if and only if the ordinary sequence $(z_{m,j}, \leq)$ has the interpolation property. Making use of result *a)* we obtain: *b) A double sequence* $(z_{m,j})$ *in* Ω *of distinct elements has the interpolation property if and only if, for an arbitrary compact subset* K *of* Ω, *there is* $(m_0, j_0) \in \mathbb{N} \times \mathbb{N}$ *such that* $z_{m,j} \notin \hat{K}$, *for* $(m,j) \geq (m_0, j_0)$.

If (z_j) is a sequence in Ω with the interpolation property and we write $Tf = (f(z_j))$, $f \in \mathcal{H}(\Omega)$, we then have that

$$T: \ \mathcal{H}(\Omega) \longrightarrow \omega$$

is a linear map which is continuous and onto, therefore it is a topological homomorphism. Let us assume now that set of adherent points of (z_j) coincides with the boundary $\partial\Omega$. Let F be the subspace of $\mathcal{H}(\Omega)$ formed by all those functions f which are zero in z_j, $j \in \mathbb{N}$, and let G be the subspace of $\mathcal{H}(\Omega)$ formed by those f such that $f(z_j) = 0$ except for a finite number of subindexes j. Clearly, the elements of F, and also those of G, which are different from the zero-function, do not extend holomorphically outside Ω. It follows that F is the kernel of T and, since the topology of $\mathcal{H}(\Omega)$ is defined by a family of norms, $\mathcal{H}(\Omega)$ is not isomorphic to ω, and so F has infinite dimension. On the other hand, $T(G)$ is the subspace of ω whose elements are the complex sequences whose terms are all zero except for a finite number. Since this subspace is dense in ω and F is contained in G, it follows that $G = T^{-1}(T(G))$ is dense in $\mathcal{H}(\Omega)$.

In [1], a domain Ω of $\mathbb{C}$ and a sequence (z_j) in Ω whose set of adherent points coincides with $\partial\Omega$ are considered. For this sequence, the subspaces F and G before defined are constructed. Then, using purely holomorphic methods, it is shown there that F has infinite dimension, G is dense in $\mathcal{H}(\Omega)$ and that the non-zero elements of F, as well as those of G, do not extend holomorphically outside Ω. In this paper, there are also a number of references concerning the so-called "lineability" problems.

A locally convex space E is said to be *Baire-like* whenever, given any increasing sequence (A_j) of closed absolutely convex subsets of E covering E, then, for some j, A_j is a neighborhood of zero, [3].

A locally convex space E is said to be unordered Baire-like whenever, given any sequence (A_j) of closed absolutely convex subsets of E covering E, then, for some j, A_j is a neighborhood of zero, [4].

2. Nearly-Baire spaces

We say that a subset A of a locally convex space is sum-absorbing whenever there is $\lambda > 0$ such that $\lambda(A + A)$ is contained in A. A locally convex space E is nearly-Baire if, for any given sequence (A_j) of closed balanced sum-absorbing subsets, whose union is E, there is an index j_0 such that A_{j_0} is a zero-neighborhood. The nearly-Baire spaces belong to a broader class of locally convex spaces which we introduced in [7] related to the closed-graph theorem. Clearly, every nearly-Baire space is unordered Baire-like.

A locally convex space E is *ordered nearly-Baire* if, for any given increasing sequence (A_j) of closed balanced sum-absorbing subsets, whose union is E, there is an index j_0 such that A_{j_0} is a zero-neighborhood. Clearly, every ordered nearly-Baire space is Baire-like. In the following we shall give an example of an unordered Baire-like space that is not ordered nearly-Baire. For this matter, we take $0 < p < 1$ and we consider the subspace E of ℓ^1 defined by

$$E := \{(a_j) \in \ell^1 : \sum_{j=1}^{\infty} | a_j |^p < \infty \}.$$

We set

$$A := \{(a_j) \in E : \sum_{j=1}^{\infty} | a_j |^p \leq 1 \}, \quad A_j := jA, \quad j \in \mathbb{N}.$$

We have that (A_j) is an increasing sequence of subsets of E which covers E. Obviously, A_j is balanced. Let us now fix $j \in \mathbb{N}$ and consider (a_h) and (b_h) in A_j. Let $\lambda := 2^{1+1/p}$. Then

$$\sum_{h=1}^{\infty} | \frac{a_h + b_h}{j\lambda} |^p \leq \frac{1}{\lambda^p} \sum_{h=1}^{\infty} (| \frac{a_h}{j} | + | \frac{b_h}{j} |)^p \leq \frac{2^p}{\lambda^p} \sum_{h=1}^{\infty} (sup\{| \frac{a_h}{j} |, | \frac{b_h}{j} |\})^p$$

$$\leq \frac{2^p}{\lambda^p} (\sum_{h=1}^{\infty} (| \frac{a_h}{j} |^p + \sum_{h=1}^{\infty} | \frac{b_h}{j} |^p) \leq \frac{2^{p+1}}{\lambda^p} = 1$$

and so $(\frac{a_h+b_h}{j\lambda})_{h=1}^{\infty} \in A$ and hence $A_j + A_j \subset \lambda A_j$, thus A_j is sum-absorbing. Let $(a_j^{(r)})$, $r = 1, 2, ...$, be a sequence in A that converges to (a_j) in ℓ^1. Consequently, for each $j \in \mathbb{N}$, $\lim_r a_j^{(r)} = a_j$. Given $s \in \mathbb{N}$, we have that $\sum_{j=1}^{\infty} | a_j^{(r)} |^p \leq 1$, therefore $\sum_{j=1}^{\infty} | a_j |^p \leq 1$, and thus $(a_j) \in A$. Then, for each $j \in \mathbb{N}$, A_j is closed in ℓ^1 and it is not a neighborhood of zero in this space. It follows then that E is not ordered nearly-Baire. The space ℓ^p, with the usual metric, is a complete and metrizable locally convex space, thus being a Baire space. A fundamental system of zero-neighborhoods in ℓ^p is $\frac{1}{m}A$, $m \in \mathbb{N}$. We now show that E is unordered Baire-like. We take a sequence (B_j) of closed absolutely convex subsets of E covering E. These sets are closed in ℓ^p and they cover ℓ^p, consequently, there is $j_0 \in \mathbb{N}$ such that B_{j_0} has interior points in ℓ^p, thus being a zero-neighborhood in ℓ^p. There is thus a positive integer r for which $\frac{1}{r}A \subset B_{j_0}$. The element e_s of E such that it has the zero value for all its coordinates except for the one in position s, which has value one, belongs to A and so

$$e_s \in r\, B_{j_0}, \quad s \in \mathbb{N}.$$

The closed unit ball B of E is the closed absolutely convex hull of $\{e_s : s \in \mathbb{N}\}$ in E, therefore $B \subset rB_{j_0}$, from where we obtain that B_{j_0} is a neighborhood of the origin in E.

It is plain that every Baire space is nearly-Baire. After our coming Theorem 1, we shall give an example of a nearly-Baire space that is not Baire.

Proposition 2.1. *If E is a nearly-Baire space and F is a one-dimensional space, then $E \times F$ is nearly-Baire.*

Proof. We identify in the usual fashion E and F with subspaces of $E \times F$. Let (A_j) be a sequence of closed balanced sum-absorbing subsets of $E \times F$ whose union is $E \times F$. We prove that one of them is a zero-neighborhood. With no loss of generality, we may assume that all sets of the form rA_j, $r, j \in \mathbb{N}$ (in the sequel we shall refer to these sets as the integral homotetics of (A_j)), belong to the sequence (A_j). Let (B_j) be the subsequence of (A_j) formed by all those A_j which meet F in more than one point. We take $(0, a) \in F$, $a \neq 0$, and an arbitrary point $(x, y) \in E \times F$. The linear span G of $\{(0, a), (x, y)\}$ clearly has finite dimension and hence there is $j_0 \in \mathbb{N}$ such that $A_{j_0} \cap G$ has non-empty interior in G, now, since A_{j_0} is balanced and sum-absorbing, it follows that $A_{j_0} \cap G$ is a neighborhood of the origin in

G. Consequently, we may find a positive integer r such that rA_{j_0} contains $\{(0,a),(x,y)\}$, thus getting that (B_j) covers $E \times F$.

We then find $j_1 \in \mathbb{N}$ such that $B_{j_1} \cap E$ be a zero-neighborhood in E. Clearly, $B_{j_1} \cap F$ is a zero-neighborhood in F. We take $\mu > 0$ such that $\mu(B_{j_1} + B_{j_1}) \subset B_{j_1}$. Let U_1 and U_2 be absolutely convex neighborhoods of zero in E and F, respectively, such that $U_1 \subset \mu B_{j_1}$ and $U_2 \subset \mu B_{j_1}$. Then, if $(x,y) \in U_1 \times U_2$, we have that

$$(x,y) = (x,0) + (0,y) \in \mu B_{j_1} + \mu B_{j_1} = \mu(B_{j_1} + B_{j_1}) \subset B_{j_1},$$

from where we obtain that B_{j_1} is a neighborhood of the origin in $E \times F$.□

For the next proposition we shall consider a locally convex space E and a subspace F of E. Let τ be a locally convex topology on F finer than the initial one and such that $F\,[\tau]$ is a nearly-Baire space. Let (A_j) be a sequence of closed balanced sum-absorbing subsets of E covering E, such that the integral homotetics of (A_j) are also in the sequence.

Proposition 2.2. *Let (B_j) be the subsequence of (A_j) formed by all A_j that intersect F in a τ-neighborhood of the origin. Then $\cup_{j=1}^{\infty} B_j = E$.*

Proof. On assuming that $\cup_{j=1}^{\infty} B_j \neq E$, there is a one-dimensional subspace L of E such that $B_j \cap L = \{0\}$, $j \in \mathbb{N}$. Let $H := F + L$, and let $\mathcal{U}$ be the locally convex topology in H such that $H\,[\mathcal{U}]$ is isomorphic to $F[\tau] \times L$. After our former proposition, $H[\mathcal{U}]$ is nearly-Baire. Thus, there is a positive integer j_0 such that $A_{j_0} \cap H$ is a zero-neighborhood in $H[\mathcal{U}]$. Hence, A_{j_0} coincides with some B_j and meets L in more than one point, which is a contradiction. □

Proposition 2.3. *If E and F are nearly-Baire spaces, then $E \times F$ is also nearly-Baire.*

Proof. Let (A_j) be a sequence of closed balanced sum-absorbing subsets which cover $E \times F$. Our goal is to show that there is some $j \in \mathbb{N}$ such that $A_j$is a zero-neighborhood, we may thus assume that (A_j) contains its integral homotetics. Let (B_j) be the subsequence of (A_j) formed by those A_j which meet F in a zero-neighborhood. By applying the former proposition, we obtain that $\cup_{j=1}^{\infty} B_j = E \times F$. Let (C_j) be the subsequence of (B_j) formed by those B_j meeting E in a zero-neighborhood. Again our former proposition yields $\cup_{j=1}^{\infty} C_j = E \times F$. Thus every C_j meets E and F in a zero-neighborhood. Now, following a similar argument to the one used in the proof of Proposition 2.1, it can be seen that C_j is neighborhood of zero in $E \times F$. □

For an uncountable set I, we showed in [6] that in the Banach space $c_0(I)$ there are two dense subspaces X_1 and X_2 which are Baire but whose product $X_1 \times X_2$ is not. Clearly, X_1 and X_2 are nearly-Baire and, after Proposition 2.3, $X_1 \times X_2$ is also nearly-Baire. We thus have that $X_1 \times X_2$ is a nearly-Baire space that is not Baire.

Let I be a nonempty set. For each i in I, let E_i be a locally convex space. Let $E := \prod_{i \in I} E_i$. We consider, in the usual way, E_i as a subspace of E. If M is a nonempty subset of I, we identify, in the usual way, $\prod_{i \in M} E_i$ with a subspace of E.

Theorem 2.1. *Every product of ordered nearly-Baire spaces is an ordered nearly-Baire space.*

Proof. We take first a finite number $E_1, E_2, ..., E_m$ of ordered nearly-Baire spaces. Let A be a closed balanced sum-absorbing subset of $\prod_{r=1}^{m} E_r$ such that $A \cap E_r$ is a neighborhood of the origin in E_r, $r = 1, 2, ..., m$. Then, $A \cap (E_1 \times E_2)$ is a closed balanced sum-absorbing subset of $E_1 \times E_2$ such that $(A \cap (E_1 \times E_2) \cap E_1 = A \cap E_1$ is a neighborhood of the origin in E_1 and $(A \cap (E_1 \times E_2) \cap E_2 = A \cap E_2$ is a neighborhood of the origin in E_2, therefore, making use of part of the proof of Proposition 2.1, it follows that $A \cap (E_1 \times E_2)$ is a neighborhood of the origin in $E_\times E_2$. We thus have that $A \cap (E_1 \times E_2)$ is a neighborhood of the origin in $E_\times E_2$ and $A \cap E_3$ is a neighborhood of the origin in E_3, so the same argument used before yields that $A \cap (E_1 \times E_2 \times E_3)$ is a zero-neighborhood in $E_1 \times E_2 \times E_3$. Proceeding in this manner, we obtain that $A \cap (E_1 \times E_2 \times ... \times E_m) = A$ is a zero-neighborhood in $E_1 \times E_2 \times ... \times E_m$.

Let us now take a nonempty subset I and, for each i in I, an ordered nearly-Baire space E_i. Let (A_j) be an increasing sequence of closed balanced sum-absorbing subsets of $E := \prod_{i \in I} E_i$ covering E. Let us assume first that I is finite, $I = \{i_1, i_2, ..., i_m\}$. For $1 \leq r \leq m$, we have that $(A_j \cap E_{i_r})$ is an increasing sequence of closed balanced sum-absorbing subsets of E_{i_r} such that they cover E_{i_r} and so there is a positive integer j_r such that $A_{j_r} \cap E_{i_r}$ is a neighborhood of the origin in E_{i_r}, $r = 1, 2, ..., m$. We take an integer s bigger than $j_1, j_2, ..., j_m$. Then, $A_s \cap E_{i_r}$ is a zero-neighborhood in E_{i_r}, $r = 1, 2, ..., m$, and, from what was said before, $A_s \cap E$ is a zero-neighborhood in E, from where we deduce that E is an ordered nearly-Baire space. Let us now assume that I is infinite. We show that there are a finite subset M of I and a positive integer r such that

$$A_r \supset E_i, \quad i \in I \setminus M.$$

On assuming the property is not true, we find a sequence (i_j) in I of distinct terms such that

$$E_{i_j} \not\subset A_j, \quad j \in \mathbb{N}.$$

We take $x_j \in E_{i_j} \setminus A_j$. If L_j is the linear span of $\{x_j\}$ and L denotes the closed linear span in E of $\{x_j : j \in \mathbb{N}\}$, we have that L is isomorphic to the Fréchet space $\prod_{j\in\mathbb{N}} L_j$ and the sequence $(A_j \cap L)$ covers L. Consequently, there is j_0 in $\mathbb{N}$ such that $A_{j_0} \cap L$ is a zero-neighborhood in L. Hence, A_{j_0} contains all the subspaces L_j except at most a finite number $L_{j_1}, L_{j_2}, ..., L_{j_p}$ of them. If we now take an integer $j_1 > j_0$ such that $x_{j_s} \in A_{j_1}$, $s = 1, 2, ..., p$, then $x_n \in A_{j_1}$, $n \in \mathbb{N}$, which is a contradiction. Thus, there exist a finite subset M of I and a positive integer r such that

$$E_i \subset A_r, \quad i \in I \setminus M.$$

Now, since A_r is closed and sum-absorbing, we have that

$$\prod_{i\in I\setminus M} E_i \subset A_r.$$

Having in mind that E_i is an ordered nearly-Baire space, we may find an integer $m > r$ such that $A_m \cap E_i$ is a zero-neighborhood in E_i for each i in M. Then, from what was said above, $A \cap \prod_{i\in M} E_i$ is a zero-neighborhood in $\prod_{i\in M} E_i$. Thus, A_r is a neighborhood of the origin in E and so E is ordered nearly-Baire. □

Theorem 2.2. *Let E be an ordered nearly-Baire space. If F is a subspace of E of countable codimension, then F is ordered nearly-Baire.*

Proof. Let (A_j) be a sequence of closed balanced sum-absorbing subsets of F which covers F. Let $\overline{A_j}$ be the closure of A_j in E, $j \in \mathbb{N}$. Let us assume first that F is a hyperplane of E. If, for some $j_0 \in \mathbb{N}$, $\overline{A_{j_0}} \neq A_{j_0}$, then $\cup_{m=1}^{\infty} m\overline{A_{j_0}} = E$ and so $\overline{A_{j_0}}$ is a zero-neighborhood in E, thus yielding that $A_{j_0} = \overline{A_{j_0}} \cap F$ is a zero-neighborhood in F. If the linear span of $\overline{A_j}$ is F, $j \in \mathbb{N}$, we take x in $E \setminus F$ and we put

$$M_j := \{\lambda x : |\lambda| \leq j\}.$$

Then, $(A_j + M_j)$ is an increasing sequence of closed balanced sum-absorbing subsets of E such that it covers E. Consequently, there is a positive integer j_0 such that $A_{j_0} + M_{j_0}$ is a zero-neighborhood in E and so

$A_{j_0} = (A_{j_0} + M_{j_0}) \cap F$ is a zero-neighborhood in F. We may conclude from here that, if F has finite codimension in E, then F is an ordered nearly-Baire space. Let us now assume that the codimension of F in E is infinite. Let G be the closed linear hull of $\cup_{j=1}^{\infty} \overline{A_j}$. If the codimension of G in E is finite, we have that G is ordered nearly-Baire and, since obviously $\cup_{j=1}^{\infty} j\overline{A_j} = G$, there is $j_0 \in \mathbb{N}$ such that $j_0\overline{A_{j_0}}$ is a zero-neighborhood in G, and so $A_{j_0} = \overline{A_{j_0}} \cap F$ is a zero-neighborhood in F. If the codimension of G in E is infinite, let $\{x_j : j \in \mathbb{N}\}$ be a cobasis of G in E. For each $j \in \mathbb{N}$, we write

$$M_j := \{\lambda_1 x_1 + \lambda_2 x_2 + ... + \lambda_j x_j : |\lambda_i| \leq j, i = 1, 2, ..., j\}.$$

Then, $(j\overline{A_j} + M_j)$ is an increasing sequence of closed balanced sum-absorbing subsets of E covering E, hence there is $j_0 \in \mathbb{N}$ such that $j_0\overline{A_{j_0}} + M_{j_0}$ is a zero-neighborhood in E. Thus, since

$$j_0 A_{j_0} = (j_0\overline{A_{j_0}} + M_{j_0}) \cap F,$$

it follows that A_{j_0} is a zero-neighborhood in F. Therefore, F is an ordered nearly-Baire space. □

The general statement for a so called three-space-problem reads as follows: *Let X be a locally convex space and let L be a closed linear subspace of X. If L and X/L enjoy a certain property $\mathcal{P}$, will this imply that X also enjoys this property $\mathcal{P}$?* For some properties $\mathcal{P}$ this problem has a straightforward solution. We shall prove in the following that, if $\mathcal{P}$ is the property of "being nearly-Baire", then the corresponding three-space problem has a positive answer. In order to do so, some previous preparation is required.

In the next proposition we assume that F is a closed linear subspace of a locally convex space E and that T is the canonical mapping from E onto E/F. A will be a closed balanced sum-absorbing subset of E. The idea of the proof of this proposition relies mainly on the methods used in [2].

Proposition 2.4. *Let U be an absolutely convex zero-neighborhood in E, such that $U \cap F \subset A$. If $\overline{T(A \cap U)}$ is a zero-neighborhood in E/F, then A is a zero-neighborhood in E.*

Proof. We take $\lambda > 0$ such that $\lambda(A + A) \subset A$. The set

$$T(E \setminus \overline{\frac{1}{3}\lambda(A \cap U) + F})$$

is an open subset of E/F disjoint with $T(\frac{1}{3}\lambda(A \cap U) + F) = \frac{1}{3}\lambda T(A \cap U)$. Thus, $\frac{1}{3}\lambda T^{-1}(\overline{T(A \cap U)})$ is contained in $\overline{\frac{1}{3}\lambda(A \cap U) + F}$, and so this set is a zero-neighborhood in E. Consequently, $W := \frac{1}{3}\lambda U \cap \overline{\frac{1}{3}\lambda(A \cap U) + F}$ is a zero-neighborhood in E. Now, if $\mathcal{V}$ is a fundamental system of zero-neighborhoods in E, we have that

$$\overline{\frac{1}{3}\lambda(A \cap U) + F} = \bigcap_{V \in \mathcal{V}} (V + \frac{1}{3}\lambda(A \cap U) + F) =$$

$$\bigcap_{V \in \mathcal{V}} (\frac{1}{3}\lambda(V \cap U) + \frac{1}{3}\lambda(A \cap U) + F).$$

If we fix V in $\mathcal{V}$ and x in W, it follows that

$$x \in \frac{1}{3}\lambda U, \quad x = x_1 + x_2 + x_3, \ x_1 \in \frac{1}{3}\lambda(V \cap U), \quad x_2 \in \frac{1}{3}\lambda(A \cap U), \quad x_3 \in F,$$

and so

$$x_3 = x - x_1 - x_2 \in \frac{1}{3}\lambda U + \frac{1}{3}\lambda U + \frac{1}{3}\lambda U = \lambda U.$$

Therefore

$$x \in \frac{1}{3}\lambda(V \cap U) + \frac{1}{3}\lambda(A \cap U) + \lambda(F \cap U),$$

from where we get

$$W \subset \bigcap_{V \in \mathcal{V}} (\frac{1}{3}\lambda(V \cap U) + \frac{1}{3}\lambda(A \cap U) + \lambda(F \cap U)) = \overline{\frac{1}{3}\lambda(A \cap U) + \lambda(F \cap U)}$$

$$\subset \overline{\frac{1}{3}\lambda(A \cap U) + \lambda A} \subset \overline{\lambda A + \lambda A} \subset \overline{A} = A.$$

Thus A is a neighborhood of zero in E. □

Theorem 2.3. *Let F be a closed linear subspace of a locally convex space E. If F and E/F are ordered nearly-Baire, then E is also ordered nearly-Baire.*

Proof. Let (A_j) be an increasing sequence of closed balanced sum-absorbing subsets of E which covers E. Then, $(A_j \cap F)$ is an increasing sequence of closed balanced sum-absorbing subsets of F covering F and so there is an index r such that $A_r \cap F$ is a zero-neighborhood in F. We may now assume, choosing a subsequence if necessary, that $A_1 \cap F$ is a zero-neighborhood in F and $A_j \supset jA_1$, $j \in \mathbb{N}$. We take a closed absolutely convex neighborhood of the origin U in E such that $F \cap U$ is contained in A_1. The sequence $(A_j \cap jU)$ is formed by balanced sum-absorbing subsets of E covering E. Thus, if T denotes the canonical map from E onto E/F, $\overline{T(A_j \cap jU)}$ is an increasing sequence of closed balanced sum-absorbing subsets of E/F which cover E/F. Hence, there is a positive integer s such that $\overline{T(A_s \cap sU)}$ is a zero-neighborhood in E/F. On the other hand,

$$(jU) \cap F \;=\; j(U \cap F) \;\subset\; jA_1 \;\subset\; A_j$$

and so, applying Proposition 2.4, we obtain that A_j is a zero-neighborhood in E. Therefore, E is an ordered nearly-Baire space □

Theorem 2.4. *Let F be a closed linear subspace of a locally convex space E. If F and E/F are nearly-Baire spaces, then E is also nearly-Baire.*

Proof. Let (A_j) be a sequence of closed balanced and sum-absorbing subsets of E covering E. We want to find an index j such that A_j is a neighborhood of zero in E, and therefore it means no loss of generality to assume as before that the integral homotetics of each A_j are all contained in the sequence. After Proposition 2, we may also assume that $A_j \cap F$ is a zero-neighborhood in F, $j \in \mathbb{N}$. For each $j \in \mathbb{N}$, we find a closed absolutcly convex zero-neighborhood U_j in E such that $F \cap U_j \subset A_j$. We now consider the collection of sets

$$\{\, mA_j \cap U_j \;:\; m, j \in \mathbb{N}\}.$$

This collection covers E and each of its members is closed balanced and sum-absorbing. Hence the sets in

$$\{\, \overline{T(mA_j \cap U_j)} :\; m, j \in \mathbb{N}\}$$

are closed balanced sum-absorbing and they cover E/F. Since this quotient is nearly-Baire, it follows that there exist $m_0, j_0 \in \mathbb{N}$ such that $\overline{T(m_0 A_{j_0} \cap U_{j_0})}$ is a neighborhood of the origin. By applying the previous proposition, we obtain that A_{j_0} is a zero-neighborhood in E. □

3. The space $\mathcal{H}(\Omega)$

In this section, we fix the domain Ω of $\mathbb{C}^n$, $n \in \mathbb{N}$, and denote by $\mathcal{K}$ the collection of all compact subsets of Ω. For $K \in \mathcal{K}$, we write $L(K)$ to denote the intersection with $\partial\Omega$ of the closure of $\hat{K}$ in $\mathbb{C}^n$. We shall use $\partial_0\Omega$ and $\partial_s\Omega$ to represent the closure and interior in $\partial\Omega$, respectively, of

$$\partial\Omega \setminus \cup\{L(K) :\; K \in \mathcal{K}\}.$$

In all the following, except in the corollary after Theorem 3, we shall assume that $\partial_s\Omega \neq \emptyset$.

Given an open subset A in $\partial\Omega$, $z_0 \in A$ and $f \in \mathcal{H}(\Omega)$, we say that f extends holomorphically in z_0 through A whenever there is a ball $B(z_0, r)$, not meeting $\partial\Omega \setminus A$, and there is a function g holomorphic in $B(z_0, r)$ which coincides with f in a connected component of $B(z_0, r) \cap \Omega$.

We shall say that an open set A in $\partial\Omega$ is completely singular if, given an arbitrary point z_0 in A, there is a function f of $\mathcal{H}(\Omega)$ that does not extend holomorphically in z_0 through A. We shall say that a function g of $\mathcal{H}(\Omega)$ does not extend holomorphically outside Ω through A whenever it does not extend holomorphically in every point z_0 of A through A.

Proposition 3.1. *There exists a sequence (z_j) in Ω whose set of adherent points is $\partial_0\Omega$ such that, if f is in $\mathcal{H}(\Omega)$ and $\lim_j \mid f(z_j) \mid = \infty$, then f does not extend holomorphically outside Ω through $\partial_s\Omega$.*

Proof. We shall say that an element z of $\mathbb{C}^n$ is *rational* whenever the real part and the imaginary part of each one of its coordinates are rational numbers. We consider all the polydisks $B(z, \rho)$, with z and ρ rational, such that z is in Ω and

$$B(z, \rho) \cap \partial\Omega \ = \ B(z, \rho) \cap \partial_s\Omega \ \neq \ \emptyset.$$

We put $A(z, \rho)$ to denote the connected part of $B(z, \rho) \cap \Omega$ to which z belongs. Since $A(z, \rho)$ does not coincide $B(z, \rho)$, there are points in the boundary of $A(z, \rho)$ which belong to $B(z, \rho) \cap \partial_s\Omega$ and therefore, given an arbitrary compact subset K of Ω, there are points of $A(z, \rho)$ which are not in $\hat{K}$. We take now in Ω a fundamental system of compact sets

$$K_1 \subset K_2 \subset ... \subset K_j \subset ...$$

In $\partial\Omega \setminus \cup\{L(K) \ : \ K \in \mathcal{K}\}$, we take a dense subset $\{u_j \ : \ j \in \mathbb{N}\}$. For each (j, k) in $\mathbb{N} \times \mathbb{N}$, we choose $r_{j,k} > 0$ being less than both $\frac{1}{2^{j+k}}$ and the distance $d(u_j, \hat{K}_{j+k})$ from u_j to $\hat{K}_{j+k}$. The sets $B(u_j, \hat{K}_{j+k})$, $j, k \in \mathbb{N}$, and $B(z, \rho)$, when z and ρ vary with the conditions above imposed, may be ordered so that they form the sequence (A_j).

We now take $u_{1,1}$ in $A_1 \setminus \hat{K}_1$. Proceeding inductively, we assume to have found, for a positive integer m, the elements $u_{j,k}$ in Ω such that $2 \leq j + k \leq m + 1$. We take in $A_j \setminus \hat{K}_{m+1}$ the elements $u_{j,m+2-j}$, $j = 1, 2, ..., m + 1$, such that they are all distinct and different to the ones previously obtained. This concludes the complete induction process.

We thus have that $(u_{j,k}, \leq)$, where $\leq$ is the order relation in $\mathbb{N} \times \mathbb{N}$ introduced before, is a sequence in Ω which may be easily seen to have the interpolation property and whose set of adherent points coincides with $\partial_0\Omega$. We write (z_j) to denote this sequence. We then take f in $\mathcal{H}(\Omega)$ such that

$$\lim_j \mid f(z_j) \mid = \infty.$$

Let us suppose that f can be extended holomorphically in a point z_0 of $\partial_s\Omega$. Then, there are $r > 0$ and a holomorphic function g in $B(z_0, r)$ such that $B(z_0, r)$ does not intersect $\partial\Omega \setminus \partial_s\Omega$, and there is a connected component A of $B(z_0, r) \cap \Omega$ in which f and g coincide. We take $v_1 \in \partial A$, $v_1 \in B(z_0, r)$. We choose $r_1 > 0$ so that $B(v_1, 2r_1)$ is contained in $B(z_0, r)$. We take a rational point $z \in B(v_1, r_1) \cap A$. Let r_2 be the distance from z to $\partial\Omega$. Then, $r_2 \leq \| z - v_1 \| < r_1$ and so we can take ρ being rational and such that $r_2 < \rho < r_1$. If $w \in \mathbb{C}^n$ is such that $\| w - z \| \leq \rho$, we have that

$$\| w - v_1 \| \leq \| w - z \| + \| z - v_1 \| < \rho + r_1 < 2r_1$$

and thus $\overline{B(z, \rho)}$ is contained in $B(z_0, r)$, from where we deduce that g is bounded in $B(z, \rho)$. There is j_0 in $\mathbb{N}$ such that the connected component of $B(z, \rho) \cap \Omega$ to which z belongs coincides with A_{j_0}. Since A_{j_0} contains infinitely many terms of the sequence (z_j), we conclude that f is not bounded in A_{j_0}. On the other hand, A_{j_0} is contained in A and so g coincides with f in A_{j_0}. Noticing that g is bounded in A_{j_0}, we have achieved a contradiction.

□

Proposition 3.2. *Let $(z_{m,j})$ be a double sequence in Ω such that its terms are all distinct. If, for each $m \in \mathbb{N}$, the sequence $(z_{m,j})_{j=1}^{\infty}$ has the interpolation property, then there is a sequence of positive integers $1 = p_1 < p_2 < ... < p_r < ...$ such that, if $u_{m,j} := z_{m,p_m+j}$, $m, j \in \mathbb{N}$, then the double sequence $(u_{m,j})$ has the interpolation property.*

Proof. Let

$$K_1 \subset K_2 \subset ... \subset K_r \subset ...$$

be a fundamental system of compact subsets of Ω. We put $p_1 := 1$. Proceeding again by recurrence, let us assume that, for a positive integer m, we have found a positive integer p_m. We take now another positive integer $p_{m+1} > p_m$ such that, if $s \geq p_{m+1}$, we have $z_{m+1,s} \notin \hat{K}_{m+1}$. We set $u_{m,j} := z_{m,p_m+j}$, $m, j \in \mathbb{N}$, and we show that $(u_{m,j})$ has the interpolation property. Let $K \in \mathcal{K}$. We find a positive integer m_0 such that $K \subset K_{m_0}$. Let j_0 be a positive integer such that

$$z_{m,j} \notin \hat{K}_{m_0}, \quad j \geq j_0, \quad m = 1, 2, ..., m_0.$$

We now take (m, j) in $\mathbb{N} \times \mathbb{N}$ with $(m, j) \geq (m_0, j_0)$. Then, if $m \leq m_0$, we have that $j \geq j_0$ and so $z_{m,p_m+j} \notin \hat{K}_{m_0}$, consequently $u_{m,j} \notin \hat{K}$ and, if $m > m_0$, $z_{m,p_m+j} \notin \hat{K}_m$ which yields $u_{m,j} \notin \hat{K}$. Again making use of result *b)*, we obtain that $(u_{m,j})$ has the interpolation property. □

The following theorem is a generalization of the Cartan-Thullen theorem. Its proof is included for the sake of completeness.

Theorem 3.1. *Let A be a nonempty open set in $\partial\Omega$. The following are equivalent:*

1. *A is completely singular.*
2. *A is a subset of $\partial_s\Omega$.*
3. *There is a function f in $\mathcal{H}(\Omega)$ that does not extend holomorphically outside Ω through A.*

Proof. $1 \Rightarrow 2$. Let us assume that A is not contained in $\partial_s\Omega$. Then A meets $\cup\{L(K) : K \in \mathcal{K}\}$ and so there is $K \in \mathcal{K}$ such that $A \cap L(K) \neq \emptyset$. We take $u_0 \in A \cap L(K)$. Let δ be the distance from u_0 to $\partial\Omega \setminus A$ whenever this latter set is nonempty and $\delta := 1$ if $\partial\Omega \setminus A = \emptyset$. We choose H in $\mathcal{K}$ such that K is contained in the interior $\overset{\circ}{H}$ of H. Let η be the distance from K to $\mathbb{C}^n \setminus \overset{\circ}{H}$. Let $0 < \zeta < \frac{1}{4} inf\{\eta, \delta\}$. We choose z_0 in $B(u_0, \zeta) \cap \hat{K}$. Let f be an arbitrary element of $\mathcal{H}(\Omega)$. Given $\alpha \in \mathbb{N}_0^n$, we find $z_\alpha \in K$ such that

$$| D^\alpha f(z_\alpha) | = \ sup\{| D^\alpha f(z) | : \ z \in K\}.$$

The function f is holomorphic in $B(z_\alpha, \eta)$, therefore, applying Cauchy's formula, there is a constant $M > 0$ such that

$$| D^\alpha f(z_\alpha) | \leq \frac{M\alpha!}{(4\zeta)^{|\alpha|}}.$$

If $z \in B(z_0, 2\zeta)$, it follows that

$$| \frac{D^\alpha f(z_0)}{\alpha!}(z - z_0)^\alpha | \leq \frac{| D^\alpha f(z_0) |}{\alpha!}(2\zeta)^{|\alpha|}$$

$$\leq \frac{| D^\alpha f(z_\alpha) |}{\alpha!}(2\zeta)^{|\alpha|} \leq \frac{M}{(4\zeta)^{|\alpha|}}(2\zeta)^{|\alpha|} = \frac{M}{2^{|\alpha|}}.$$

Since the series $\sum_{\alpha \in \mathbb{N}_0^n} 1/2^{|\alpha|}$ is convergent, we have that

$$\sum_{\alpha \in \mathbb{N}_0^n} \frac{D^\alpha f(z_0)}{\alpha!}(z - z_0)^\alpha$$

represents a function h which is holomorphic in $B(z_0, 2\zeta)$. We have $z_0 \in B(u_0, \zeta)$ and thus $B(u_0, \zeta) \subset B(z_0, 2\zeta)$. Consequently, if g is the restriction of h to $B(u_0, \zeta)$, then f coincides with g in the connected component of $B(u_0, \zeta) \cap \Omega$, to which z_0 belongs, which is a contradiction. $2 \Rightarrow 3$. After Proposition 3.1 , we may take a sequence (z_j) in Ω such that its set of adherent points coincides with $\partial_0\Omega$ and such that it has the interpolation property so that, if $f \in \mathcal{H}(\Omega)$ is such that $f(z_j) = j$, $j \in \mathbb{N}$, then f satisfies the required property. $3 \Rightarrow 1$. Evident. □

Corollary 3.1. *(Cartan-Thullen). A domain Ω in $\mathbb{C}^n$ is a domain of holomorphy if and only if it is holomorphically convex.*

We shall give next an interpolation theorem which is related with functions of $\mathcal{H}(\Omega)$ that cannot be extended holomorphically outside Ω through $\partial_s\Omega$. But, before doing this, we make the following construction:

We consider a double sequence $(z_{j,k})$ in Ω with the interpolation property. We choose a double sequence $(a_{j,k})$ of complex numbers. Let $K \in \mathcal{K}$, $\overset{\circ}{K} \neq \emptyset$, and a subset $\{w_j : j \in \mathbb{N}\}$ of K, dense in K, whose elements are all different from $z_{j,k}$, $j, k \in \mathbb{N}$. We get $f_1 \in \mathcal{H}(\Omega)$ such that, for each $k \in \mathbb{N}$,

$$f_1(z_{1,k}) = a_{1,k}, \quad f_1(z_{2,k}) = 2 + k, \quad f_1(z_{j,k}) = 0, \quad j = 3, 4, ...$$

We also find $f_2 \in \mathcal{H}(\Omega)$ such that $f_2(w_1) = f_1(w_1)$ and, for each $k \in \mathbb{N}$,

$$f_2(z_{1,k}) = a_{2,k}, \quad f_2(z_{j,k}) = j + k, \quad j = 2, 3, \quad f_2(z_{j,k}) = 0, \quad j = 4, 5, ...$$

Proceeding by recurrence, and assuming that for a positive integer $m > 1$ we have already found f_j in $\mathcal{H}(\Omega)$, $1 \leq j \leq m$, we find a function f_{m+1} in $\mathcal{H}(\Omega)$ such that $f_{m+1}(w_j) = f_j(w_j)$, $j = 1, 2, ..., m$, and, for each $k \in \mathbb{N}$,

$$f_{m+1}(z_{1,k}) = a_{m+1,k}, \quad f_{m+1}(z_{j,k}) = j + k, \quad j = 2, 3, ..., m + 2,$$

$$f_{m+1}(z_{j,k}) = 0, \quad j = m + 3, m + 4, ...$$

This completes the induction process.

Theorem 3.2. *Let (z_j) be a sequence in Ω whose set of adherent points coincides with $\partial_0\Omega$ and having the interpolation property. For a given double sequence $(a_{j,k})$ of complex numbers, there is a sequence (f_j) in $\mathcal{H}(\Omega)$ with the following properties:*

1. *$f_1, f_2, ..., f_j, ...$ are linearly independent.*
2. *For $j \in \mathbb{N}$, $f_j(z_k) = a_{j,k}$, $k \in \mathbb{N}$.*
3. *If F is the closed linear span of $\{f_j : j \in \mathbb{N}\}$ in $\mathcal{H}(\Omega)$, we have that every $f \in F$, $f \neq 0$, does not extend holomorphically outside Ω through $\partial_s\Omega$.*

Proof. We set $z_{1,j} := z_j$, $j \in \mathbb{N}$. Making use of Proposition 3.1, we may construct, for each $j \in \mathbb{N} \setminus \{1\}$, a sequence of points in Ω, $(z_{j,k})_{k=1}^{\infty}$, with the properties there mentioned and such that, using Proposition 3.2, the double sequence $(z_{j,k})$ has the interpolation property. We take a compact subset K of Ω with $\overset{\circ}{K} \neq \emptyset$, and choose a dense subset $\{w_j : j \in \mathbb{N}\}$ of K, whose elements are all different from $z_{j,k}$, $j, k \in \mathbb{N}$. By the construction before introduced, we determine the functions $f_1, f_2, ..., f_j, ...$. It is plain that these functions satisfy 1. and 2. We see next that 3. is also satisfied. Let us take a function f of F which extends holomorphically in a point z_0 of $\partial_s\Omega$ through $\partial_s\Omega$, we show that f is identically zero. For each $j \in \mathbb{N}$, we find a sequence of complex numbers $(\lambda_{j,h})$, where all terms are zero from a certain h on, so that the sequence $(\sum_{h=1}^{\infty} \lambda_{j,h} f_h)_{j=1}^{\infty}$ converges to f in $\mathcal{H}(\Omega)$. Then

$$f(z_{2,k}) = \lim_j \sum_{h=1}^{\infty} \lambda_{j,h} f_h(z_{2,k}) = (2+k) \lim_j \sum_{h=1}^{\infty} \lambda_{j,h} = (2+k)\eta.$$

If η were non-zero, then

$$\lim_k \mid f(z_{2,k}) \mid = \infty,$$

and, since the set of adherent points of $(z_{2,k})$ is $\partial_0\Omega$, we would have that f does not extend holmorphically outside Ω through $\partial_s\Omega$. Thus, $\eta = 0$. Let now $r > 2$. We have

$$f(z_{r,k}) = \lim_j \sum_{h=1}^{\infty} \lambda_{j,h} f_h(z_{r,k}) = (r+k) \lim_j \sum_{h=r-1}^{\infty} \lambda_{j,h} = (r+k)\zeta,$$

and proceeding as before, $\zeta = 0$. Consequently,

$$\lim_j \lambda_{j,h} = 0, \quad h \in \mathbb{N}.$$

It follows that, for each $s \in \mathbb{N}$,

$$f(w_s) = \lim_j \sum_{h=1}^{\infty} \lambda_{j,h} f_h(w_s) = \lim_j \sum_{h=1}^{s-1} \lambda_{j,h} f_h(w_s) + f_s(w_s) \lim_j \sum_{h=s}^{\infty} \lambda_{j,h} = 0,$$

and so, $f = 0$. □

Using the Continuum Hypothesis, we shall prove in Theorem 3.3, the existence of a linear subspace F of $\mathcal{H}(\Omega)$, dense and ordered nearly-Baire, such that, if $f \in F$, $f \neq 0$, then f does not extend holomorphically outside Ω through $\partial_s \Omega$. We dedicate some of the following paragraphs to make some more previous constructions.

Let $\mathcal{P}$ represent the collection of all sequences (P_j) whose terms are finite subsets of $\mathbb{N}$, so that they form a partition of $\mathbb{N}$ and the elements in P_{j+1} are all posterior to those of P_j, $j \in \mathbb{N}$. Let X be the set of all the sequences of complex numbers whose terms are distinct from zero.

Under the Continuum Hypothesis, the elements $p \in X \times \mathcal{P}$ may be written as p_α, $0 \leq \alpha < \omega_1$, where ω_1 denotes the first uncountable ordinal. Given $0 < \beta < \omega_1$, we take a mapping f_β from $\mathbb{N}$ onto the ordinal interval $[0, \beta[$.

For a given pair $p_0 := ((a_j), (P_j))$, we find a sequence (ρ_k) of positive integers such that it diverges to ∞ and, for each $j \in \mathbb{N}$ and $l_1, l_2 \in P_j$, it happens that $\rho_{l_1} = \rho_{l_2}$ and then

$$\lim_k \rho_k \mid a_k \mid \; = \; \infty.$$

We write $b_j^{(0)} := \rho_j a_j$, $j \in \mathbb{N}$. Proceeding by transfinite induction, let us assume we have $0 < \beta < \omega_1$ so that, for each ordinal γ, with $0 \leq \gamma < \beta$, we have already obtained the sequence $(b_j^{(\gamma)})$ of complex numbers. For every $r, j \in \mathbb{N}$, we set $c_j^{(r)} := b_j^{(f_\beta(r))}$.

If $p_\beta = ((g_j), (Q_j))$, we find a sequence (η_k) of positive integers such that it diverges to ∞ and, for each $j \in \mathbb{N}$, $l_1, l_2 \in Q_j$, we have $\eta_{l_1} = \eta_{l_2}$, and

$$\lim_k \eta_k \mid g_k \mid = \infty$$

and, for each k,

$$\eta_k \mid g_k \mid > \; k(1 + \mid c_k^{(1)} \mid + \mid c_k^{(2)} \mid + ... + \mid c_k^{(k)} \mid).$$

We write $b_j^{(\beta)} := \eta_j g_j$, $j \in \mathbb{N}$. This completes the transfinite induction process. We put e_r for the sequence whose terms are all zero except the r-coordinate which has value one. We define H as the linear span of

$$\{(b_j^{(\alpha)}) : \; 0 \leq \alpha < \omega_1\} \cup \{e_r : \; r \in \mathbb{N}\}.$$

For every $r \in \mathbb{N}$, let H_r be the subspace of H given by the closed linear hull of $\{e_r, e_{r+1}, ..., \}$, and, if $r \leq s$, $s \in \mathbb{N}$, let $H_{r,s}$ be the linear span of $\{e_r, e_{r+1}, ..., e_s\}$, finally, let K_r be the linear span of $\{e_1, e_2, ..., e_r\}$ and K_0 the null subspace. In H, $p_{r,s}$ is the projection onto $H_{r,s}$ along $K_{r-1} + H_{s+1}$.

Proposition 3.3. *In H, let (A_j) be an increasing sequence of closed balanced sum-absorbing subsets covering H. Then, there are $r, m \in \mathbb{N}$ such that $\overline{p_{r,s}(A_m)} = H_{r,s}$, for $s \in \mathbb{N}$, $s \geq r$.*

Proof. With no loss of generality, we may assume that the integral homotetics of the sequence (A_j) are contained in the sequence. Suppose now that the property stated were not true. We find a first positive integer r_1 such that

$$\overline{p_{1,r_1}(A_1)} \neq H_{1,r_1}.$$

We set $r_0 := 0$. We now find complex numbers $c_1, c_2, ..., c_{r_1}$ such that

$$c_{r_0+1}e_{r_0+1} + c_{r_0+2}e_{r_0+2} + ... + c_{r_1}e_{r_1} \notin \overline{p_{r_0+1,r_1}(A_1)}$$

Obviously, we may take the numbers $c_{r_0+1}, c_{r_0+2}..., c_{r_1}$ disntinct from zero.

Proceeding inductively, let us assume that, for a positive integer j, we have obtained the non-negative integers $r_0 < r_1 < ... < r_j$ and the complex numbers different from zero $c_1, c_2, ..., c_{r_j}$ such that

$$c_{r_{j-1}+1}e_{r_{j-1}+1} + c_{r_{j-1}+2}e_{r_{j-1}+2} + ... + c_{r_j}e_{r_j} \notin \overline{p_{r_{j-1}+1,r_j}(A_j)}$$

Let r_{j+1} be the first integer greater than r_j such that

$$\overline{p_{r_j+1,r_{j+1}}(A_{j+1})} \neq H_{r_j+1,r_{j+1}}.$$

We choose the complex numbers distinct from zero $c_{r_j+1}, c_{r_j+2}, ..., c_{r_{j+1}}$ so that

$$c_{r_j+1}e_{r_j+1} + c_{r_j+2}e_{r_j+2} + ... + c_{r_{j+1}}e_{r_{j+1}} \notin \overline{p_{r_j+1,r_{j+1}}(A_{j+1})}.$$

This concludes the induction process.

We prove now that $\sum_{i=1}^{\infty} c_i e_i$ does not belong to H. If this were not so, we find a positive integer m such that

$$\sum_{i=1}^{\infty} c_i e_i \in A_m.$$

Then

$$\sum_{i=r_{m-1}+1}^{r_m} c_i e_i \in p_{r_{m-1}+1,r_m}(A_m) \subset \overline{p_{r_{m-1}+1,r_m}(A_m)},$$

which is a contradiction.

We consider next the partition of $\mathbb{N}$,

$$P_j := \{r_{j-1}+1, r_{j-1}+2, ..., r_j\}, \quad j \in \mathbb{N}.$$

We get an ordinal γ such that $p_\gamma = ((c_j),(P_j))$. Then, the sequence of complex numbers $(b_j^{(\gamma)}) = (\eta_j c_j)$ belongs to H, and so there is an A_l such that $(b_j^{(\gamma)})$ is in A_l. Noticing that the positive integers η_j, $j \in \mathbb{N}$, are all equal in P_l and that $p_{r_{l-1}+1,r_l}(A_l)$ is balanced, we have that

$$p_{r_{l-1}+1,r_l}((b_j^{(\gamma)})) =$$

$$\eta_{r_l}(c_{r_{l-1}+1}e_{r_{l-1}+1} + c_{r_{l-1}+2}e_{r_{l-1}+2} + \ldots + c_{r_l}e_{r_l}) \notin p_{r_{l-1}+1,r_l}(A_l),$$

which is again a contradiction, thus concluding the proof. □

Lemma 3.1. *H is an ordered nearly-Baire space.*

Proof. In H, let (A_j) be an increasing sequence of closed balanced sum-absorbing subsets which covers H. We apply the former proposition and so obtain r and m in $\mathbb{N}$ so that $\overline{p_{r,s}(A_m)} = H_{r,s}$, $s \in \mathbb{N}$, $s \geq r$. It follows that $\cup\{A_j \cap K_{r-1} : j \in \mathbb{N}\}$ coincides with K_{r-1} and, since this space has finite dimension, there is s in $\mathbb{N}$ such that $A_s \cap K_{r-1}$ is a zero-neighborhood in K_{r-1}. Since (A_j) is increasing, we may take $m > s$. Then, $A_m \cap K_{r-1}$ is a zero-neighborhood in K_{r-1}. We find $0 < \mu < 1$ such that

$$\mu(A_m + A_m) \subset A_m.$$

Let $\overline{A_m}$ denote the closure of A_m in ω. Now, if $b, c \in \mathbb{C}$ and $x, y \in \overline{A_m}$, we then have

$$bx + cy \in b\overline{A_m} + c\overline{A_m} \subset (\mid b \mid + \mid c \mid)(\overline{A_m} + \overline{A_m})$$

$$= (\mid b \mid + \mid c \mid)\mu^{-1}(\mu(\overline{A_m} + \overline{A_m})) \subset (\mid b \mid + \mid c \mid)\mu^{-1}\overline{A_m}.$$

Hence, $F := \cup_{j=1}^{\infty} j\overline{A_m}$ is a linear subspace of ω. Let us assume that A_m is not a zero-neighborhood in H. Then, $F \neq \omega$. So, we take $(a_j) \in \omega \setminus F$. We choose

$$(a_j^{(s)}) \in \mu A_m \subset A_m, \quad s \in \mathbb{N}, \quad s \geq r, \tag{1}$$

so that

$$\mid a_j^{(s)} - a_j \mid < \frac{1}{s}, \quad j = r, r+1, \ldots, s.$$

If $r = 1$, the sequence (1) converges in ω to $(a_j)) \in \overline{A_m}$, which is a contradiction. Thus, $r > 1$. Suppose next that $A_m \cap K_{r-1}$ is a compact subset of K_{r-1}. There is then a positive number k such that, if $\sum_{j=1}^{r-1} b_j e_j$ is an element of A_m, then

$$| b_j | \leq k, \quad j = 1, 2, ..., r-1. \tag{2}$$

For each $s \in \mathbb{N}$, let

$$h_s := \max\{| a_j^{(s)} |: \ j = 1, 2, ..., r-1\}$$

and let us assume that $\{h_s : s \in \mathbb{N}\}$ is not bounded. There exist then $j_0 \in \mathbb{N}$, $1 \leq j_0 < r$, and a subsequence of $(a_j^{(s)})_{s=1}^\infty$, which we shall keep denoting by $(a_j^{(s)})_{s=1}^\infty$, such that

$$0 < h_1 < h_2 < ... < h_s < ..., \quad | a_j^{(s)} | = h_s, \quad 2h_s^{-1}k < 1, \quad s \in \mathbb{N}.$$

Let us now consider the sequence

$$((\frac{2k}{h_s} a_j^{(s)})_{j=1}^\infty)_{s=1}^\infty. \tag{3}$$

It follows that

$$\lim_s \frac{2k}{h_s} a_j^{(s)} \ = \ 0, \quad j \geq r, \quad (\frac{2k}{h_s} a_j^{(s)})_{j=1}^\infty \in \mu A_m, \quad s \in \mathbb{N}.$$

Therefore, there is a subsequence of (3) which converges to an element of the form

$$\sum_{j=1}^{r-1} b_j e_j \ \in \ A_m.$$

Since

$$| b_{j_0} | = \ \lim_s \frac{2k}{h_s} | a_{j_0}^{(s)} | = \ 2k,$$

after (2), we achieve a contradiction. Thus, the set $\{h_s : s \in \mathbb{N}\}$ is bounded and so there is a subsequence of $((a_j^{(s)})_{j=1}^\infty)_{s=1}^\infty$ converging to an element of the form

$$\sum_{j=r}^\infty a_j e_j \ + \sum_{j=1}^{r-1} c_j e_j \ \in \ \overline{A_m} \ \subset \ F.$$

Now, since $(a_j) \notin F$ and, clearly, $\sum_{j=1}^{r-1}(c_j - a_j)e_j \ \in F$, it follows that

$$\sum_{j=r}^\infty a_j e_j \ + \sum_{j=1}^{r-1} c_j e_j \ - \ \sum_{j=1}^\infty a_j e_j \ = \ \sum_{j=1}^{r-1}(c_j - a_j)e_j \ \notin \ F,$$

thus obtaining a contradiction.

Let us suppose that $A_m \supset K_{r-1}$. Then, for each $s \in \mathbb{N}$,

$$- \sum_{j=1}^{r-1} a_j^{(s)} \in \mu A_m,$$

and therefore

$$\sum_{j=r}^{\infty} a_j^{(s)} e_j = \sum_{j=1}^{\infty} a_j^{(s)} e_j - \sum_{j=1}^{r-1} a_j^{(s)} e_j \in \mu A_m + \mu A_m \subset A_m.$$

Now, since the sequence $(\sum_{j=r}^{\infty} a_j^{(s)} e_j)_{s=1}^{\infty}$ converges in ω to $\sum_{j=r}^{\infty} a_j e_j$, which is not in F, another contradiction is obtained.

Finally, if $A_m \cap K_{r-1}$ contains a maximal linear subspace of dimension d, $1 \leq d < r-1$, we consider a basis $e_1^*, e_2^*, ..., e_{r-1}^*$ of K_{r-1} so that the linear span of $\{e_1^*, e_2^*, ..., e_d^*\}$ is L. Then $(a_j^{(s)})$ may be written as

$$\sum_{j=1}^{r-1} c_j^{(s)} e_j^* + \sum_{j=r}^{\infty} a_j^{(s)} e_j \in \mu A_m.$$

Now, since $- \sum_{j=1}^{d} c_j^{(s)} e_j^* \in \mu A_m$, we have that

$$\sum_{j=d+1}^{r-1} c_j^{(s)} e_j^* + \sum_{j=r}^{\infty} a_j^{(s)} e_j = \sum_{j=1}^{r-1} c_j^{(s)} e_j^* + \sum_{j=r}^{\infty} a_j^{(s)} e_j - \sum_{j=1}^{d} c_j^{(s)} e_j^*$$

$$\in \mu A_m + \mu A_m \subset A_m.$$

On the other hand, A_m intersects the linear hull of $\{e_{d+1}^*, e_{d+2}^*, ..., e_{r-1}^*\}$ in a compact set, thus a subsequence of

$$\Big(\sum_{j=d+1}^{r-1} c_j^{(s)} e_j^* + \sum_{j=r}^{\infty} a_j^{(s)} e_j\Big)_{s=1}^{\infty}$$

may be found such that it converges in ω to an element of the form

$$\sum_{j=d+1}^{r-1} c_j e_j^* + \sum_{j=r}^{\infty} a_j e_j \in \overline{A_m} \subset F.$$

Since

$$\sum_{j=d+1}^{r-1} c_j e_j^* \in F, \quad \sum_{j=1}^{r-1} a_j e_j \in F,$$

we have

$$\sum_{j=1}^{\infty} a_j e_j \;=\; (a_j) \;\in\; F,$$

thus attaining a contradiction.

Consequently, $F = \omega$ and so $\overline{A_m}$ is a zero-neighborhood in ω, from where we get that $A_m = \overline{A_m} \cap H$ is also a zero-neighborhood in H. □

The next proposition follows immediately.

Proposition 3.4. *If (b_j) is an element of the linear span of*

$$\{(b_j^{(\alpha)} : \; 0 \leq \alpha < \omega_1\},$$

then $\lim_j \mid b_j \mid = \infty$.

Theorem 3.3. *Under the Continuum Hypothesis, there exists in $\mathcal{H}(\Omega)$, $\Omega \subset \mathbb{C}$, a dense ordered nearly-Baire linear subspace G such that, if $f \in G$, $f \neq 0$, then f does not extend holomorphically outside Ω through $\partial_s \Omega$.*

Proof. We take a sequence (z_j) as in Proposition 3.1. For each $f \in \mathcal{H}(\Omega)$, we write $Tf := (f(z_j))$. Then

$$T : \; \mathcal{H}(\Omega) \longrightarrow \omega$$

is a topological homomorphism. Let H the linear subspace of ω before described and we set $G := T^{-1}(H)$. It follows that $T^{-1}(0)$ is a Fréchet space contained in G. Applying Lemma 3.1 and Theorem 2.3, we obtain that G is ordered nearly-Baire and dense in $\mathcal{H}(\Omega)$. Finally, if $f \in G$, $f \neq 0$, then, either $f(z_j) = 0$ from a certain subindex on, or, according to Proposition 3.4, the set of adherent points of the sequence $(\mid f(z_j) \mid)$ diverges to infinity. Therefore, f does not extend holomorphically outside Ω through $\partial_s \Omega$. □

Problem. Let Ω be a domain of holomorphy. Without the Continuum Hypothesis, does there exist a dense ordered nearly-Baire linear subspace G of $\mathcal{H}(\Omega)$ such that, if $f \in G$, $f \neq 0$, f does not extend holomorphically outside Ω ?

References

1. Aron, R.; Garcia, D.; Maestre, M.: Linearity in non-linear problems, *Rev. R. Acad. Cien.*, Serie A. Mat. **95 (1)**, 7-12 (2001).
2. Roelcke, W.; Dierolf, S.: On the three-space problem for topological vector spaces, *Collectanea Mathematca*, **XXXI**, 13-35 (1981).
3. Saxon, S.: Nuclear and product spaces, Baire-like spaces and the strongest locally convex topology, *Math. Ann.*, **197**, 87-106 (1972).
4. Todd, A.; Saxon, S.: A property of locally convex Baire spaces, *Math. Ann.*, **206**, 23-35 (1973).
5. Valdivia, M.: Interpolation in certain function spaces, *Proc. R. Ir. Acad.*, **80A**, 173-189 (1980).
6. Valdivia, M.: Products of Baire topological vector spaces, *Fundamenta Mathematica*, **CXXV**, 71-80 (1985).
7. Valdivia, M.: Classes of barrelled spaces related with the closed graph theorem, *Portugalia Mathematica*, **40**, 346-366 (1981).

CLASSICAL POTENTIAL THEORY AND ANALYTIC CAPACITY

JOAN VERDERA
Departament de Matemàtiques
Universitat Autònoma de Barcelona
08193 Bellaterra (Barcelona), Catalonia
E-mail: jvm@mat.uab.cat

Recently, analytic capacity has been proved to be semi-additive (that is, sub-additive modulo an absolute constant). This solves a long standing open problem in complex analysis using real variables methods. The proof of this remarkable result, which is the culmination of a long series of contributions by many authors, uses some ideas from Potential Theory and, in fact, the problem itself presents a definite formal analogy with some central aspects of Potential Theory. The purpose of this article is to review the basics of Classical Potential Theory in such a way that the formal analogy between analytic capacity and classical Wiener capacity becomes evident and the semi-additivity problem becomes natural. At the end one gives an idea of the main difficulty involved in its solution.

1. Introduction

This paper is an article version of a plenary lecture that I delivered at the "Second International Course of Mathematical Analysis ," held in Granada in the summer of 2004. In preparing the lecture I was first tempted to talk about the solution of the semi-additivity problem for analytic capacity, but I quickly realized that this would have necessarily brought the exposition into a jungle of technicalities with limited interest for the audience. Then I remembered that classical potential theory is a beautiful and powerful branch of analysis, which has been a permanent source of inspiration for many problems on analytic capacity. Therefore I planned the exposition so that it could serve as an introduction to classical potential theory for part of the audience and as an explanation of the difficulties connected with the study of analytic capacity for others. The present article follows this plan faithfully.

Section 2 contains the introduction to classical potential theory, which is inspired by [18]. Section 3 deals with analytic capacity. The interested reader is advised to read the surveys [3], [14] and [15], where further references will be found.

2. Elementary Electrostatics

2.1. *Coulomb's Law*

A charge q placed at the origin creates an electrostatic field. This means that if one places a test charge q' at the point $x \in R^3$, then a force acting on this charge appears. Coulomb's Law asserts that this force acts in the direction of the vector x (if q and q' have the same sign) and that its magnitude is proportional to the product $q\,q'$ and inversely proportional to the square of the distance between the charges q and q'. In other words, assuming for the sake of simplicity that q and q' are unit charges, the force acting on q' due to the presence of q is proportional to

$$\vec{E}(x) = \frac{x}{|x|}\frac{1}{|x|^2} = \frac{x}{|x|^3}.$$

One refers to the vector above as the electrostatic field created by a unit charge at the origin (we ignore proportionality factors). A completely analogous theory, based on Newton's Laws, can be developed when interactions due to charges are replaced by interactions due to masses, and then one talks about the gravitational field.

A simple computation shows that the electrostatic field above is the gradient of a function :

$$\frac{x}{|x|^3} = -\nabla\left(\frac{1}{|x|}\right).$$

There is an interesting interpretation of the function $\frac{1}{|x|}$ in terms of potential energy. Assume that a unit charge travels from a point x to a point y following a path $\gamma(t), a \le t \le b$. The work produced by the force $\vec{E}(x)$ is given by the line integral

$$W = \int_a^b \vec{E}(\gamma(t)) \cdot \gamma'(t)\,dt = \int_a^b -\frac{d}{dt}\left(\frac{1}{|x(t)|}\right)dt = \frac{1}{|x|} - \frac{1}{|y|}$$

Letting $y \to \infty$ one concludes that $\frac{1}{|x|}$ can be thought of as the work done by the electrostatic field when the unit charge travels from x to ∞. Alternatively, one may think that $\frac{1}{|x|}$ is the potential energy of a unit charge placed at the point x.

When the charge generating the field is placed at the point a the electrostatic field is

$$\vec{E}(x) = \frac{x-a}{|x-a|^3} = -\nabla\left(\frac{1}{|x-a|}\right)$$

and the potential energy, or simply "potential", is $\frac{1}{|x-a|}$.

Consider now a body in the space and a general charge distribution on it. In mathematical terms we are considering a compact subset K of R^3 and a signed (finite Borel regular) measure μ on K. Now the question is : what is the mathematical expression for the electrostatic field generated by the given charge distribution?. To get a quick answer we use an elegant heuristic argument very popular among physicists. Subdivide the body K into a large number of small pieces. Call q_j the total charge of the j-th piece and select a point a_j in it.

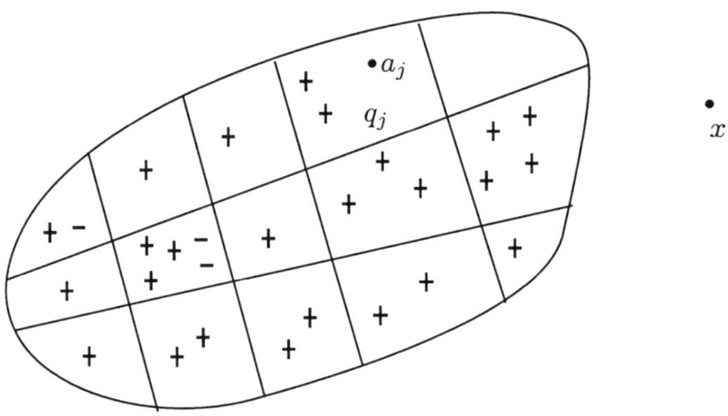

Figure 1

Then the distribution of charges μ is approximately $\sum_j q_j\, \delta_{a_j}$. Therefore the electrostatic field $\vec{E}(x)$ generated by μ is approximately the sum of the electrostatic fields generated by point charges q_j at the points a_j, that is,

$$\vec{E}(x) \simeq \sum_j q_j \frac{x-a_j}{|x-a_j|^3}.$$

Observe that the sum in the equation above is a Riemann sum associated with a certain integral. Thus, when the pieces become arbitrarily small and one takes a limit the following identity arises :

$$\vec{E}(x) = \int \frac{x-y}{|x-y|^3}\, d\mu(y).$$

In other words, the electrostatic field generated by the charge distribution μ is the convolution of the measure μ with the vector valued kernel $\frac{x}{|x|^3}$. Since in the case at hand one can take derivatives inside the integral sign, we obtain

$$\vec{E}(x) = \int \frac{x-y}{|x-y|^3}\, d\mu(y) = -\nabla\Big(\int \frac{d\mu(y)}{|x-y|}\Big).$$

The argument presented before tells us that the value

$$U^\mu(x) = \int \frac{d\mu(y)}{|x-y|}$$

represents the potential energy of a unit charge placed at the point x or the work done by the unit charge when traveling from x to ∞ under the action of the field $\vec{E}(x)$. Because of that, $U^\mu(x)$ is called the (Newtonian) potential of the measure μ. From the mathematical perspective the Newtonian potential of μ is just the convolution of μ with the positive kernel $\frac{1}{|x|}$.

We now present some basic examples.

Example 2.1. Take $\mu = \delta_0$, the Dirac delta at the origin. We are dealing with a unit charge placed at the origin and so the electrostatic field is $\vec{E}(x) = \int \frac{x}{|x|^3}\, d\delta_0(x) = \frac{x}{|x|^3}$ and the potential is $\frac{1}{|x|}$. Notice that $U^\mu(0) = \infty$, which shows that the potential energy at a point may be infinite.

Example 2.2. Let $\mu = d\sigma$ be the surface measure on the unit sphere $|x| = 1$. It is clear that the potential

$$U^\mu(x) = \int_{|y|=1} \frac{1}{|x-y|} d\sigma(y)$$

depends only on $|x|$ for $|x| > 1$. In fact one can show that

$$U^\mu(x) = \frac{4\pi}{|x|}\,, \qquad |x| > 1,$$

$$U^\mu(x) = 4\pi\,, \qquad |x| \leq 1.$$

Notice that in this case the potential is finite everywhere. This is due to the fact that the measure is much more dispersed than in the previous example, and hence is, in some sense, less singular. One obtains the electrostatic field by taking the gradient in the preceding identity:

$$\vec{E}(x) = \begin{cases} -4\pi \frac{x}{|x|^3}\,, & |x| > 1, \\ 0\,, & |x| \leq 1. \end{cases}$$

Example 2.3. Let $\mu = dt$ be the linear measure on the segment $K = \{(t,0,0): -1 \leq t \leq 1\}$. Notice that in this case the potential becomes infinite everywhere on the segment

$$U^{\mu}(x) = \int_{-1}^{1} \frac{1}{|x_1 - t|}\, dt = \infty,\ x = (x_1, 0, 0) \in K.$$

This reflects the fact that the measure is much more concentrated than in the previous example.

2.2. *The equilibrium potential*

Assume now that the body K is a conductor. What really happens when you distribute a charge, say the unit charge, on K?. The answer is that the charges move until they reach an **equilibrium distribution**. Call μ such an equilibrium charge distribution. If there were points in K with different potential energy, then this difference of potential would force the charges to move inside the conductor, which is impossible because we already reached equilibrium. Therefore $U^{\mu}(x) = V,\ \ x \in K$, where V is a constant. This is what Gauss proved assuming that the body and the potentials are as smooth as one may need. Now the following highly non-trivial mathematical problem arises.

The problem of the equilibrium distribution: *given a compact subset K of $\mathbb{R}^3$, show that there exists an "equilibrium" measure μ with total mass 1 such that the potential U^{μ} is* ***constant*** *on K.*

It turns out that an equilibrium measure exists, is unique and is a positive measure. Also, as we will see, the constant value of the "equilibrium" potential U^{μ} provides important information on K.

The problem of the equilibrium distribution was solved by Frostman, a student of Marcel Riesz, in his thesis (1935).

If the reader devotes some time to thinking about how this measure could be constructed, it becomes clear that the problem is far from easy. However, at least for the unit sphere $K = \{x : |x| = 1\}$ the equilibrium distribution can be found readily. It is the normalized surface measure. Its potential takes the constant value 1 on the sphere, as we said in Example 3 in subsection 2.1 .

The proof of the existence of an equilibrium distribution is based on the solution of an extremal problem that involves the notion of energy of a charge distribution, which we discuss in the next section.

2.3. *Energy*

Consider a signed measure μ supported on a compact subset K of $\mathbb{R}^3$. Subdivide the body K into a large number of small pieces, take a point a_j in the j-th piece and assume that the charge of the j-th piece is q_j (see Figure 1 above). Thus μ is approximately the distribution $\sum_j q_j \delta_{a_j}$. Suppose that the charge q_j goes to ∞. Then the work done by the field generated by the charges $q_k, k \neq j$ is $\sum_{k\neq j} q_k \frac{1}{|a_k - a_j|}$. When we let all charges q_j go to infinity then the total work done is

$$\sum_j q_j \sum_{k\neq j} q_k \frac{1}{|a_k - a_j|}$$

In the computation of the work just outlined we do not consider the case $k = j$ because we would immediately get the value ∞, as remarked in Example 1. Now we recognize the sum above as a Riemann sum of a certain integral. Taking the limit as the pieces become arbitrarily small, we conclude that the work done by the field generated by the distribution μ when all charges go to ∞ is exactly

$$\int U^{\mu}(x)\, d\mu(x).$$

Of course this integral should also be interpreted as the potential energy accumulated by the charge distribution μ. Hence we define the **energy** associated with the signed measure μ by the elegant expression

$$E(\mu) = \int U^{\mu}(x)\, d\mu\,(x) = \int\int \frac{d\mu(x)\, d\mu(y)}{|x-y|}.$$

Here are some examples.

If $\mu = \delta_0$, then

$$E(\delta_0) = \int \frac{1}{|x|}\, d\,\delta_0\,(x) = \infty\,.$$

If $\mu = dt$ on the segment $K = \{(t,0,0) : -1 \leq t \leq 1\}$, then

$$E(\mu) = \int_K U^{\mu}(x)\, d\mu\,(x) = \int_{-1}^{1} \infty\, dt = \infty\,.$$

The reason we get infinite energy in the two preceding examples is that the support of the charge distribution is too small. In the next two examples the charge distributions are sufficiently dispersed to yield finite energy.

Take μ to be the normalized surface measure on the unit sphere $K = \{x \in \mathbb{R}^3 : |x| = 1\}$. Since $U^\mu(x)$ is 1 on the sphere, we obtain

$$E(\mu) = \int_{|x|=1} U^\mu(x)\, d\mu(x) = 1$$

Let μ now be the volume measure (3-dimensional Lebesgue measure) restricted to the closed unit ball of $\mathbb{R}^3$. Then one can show that

$$U^\mu(x) \simeq 1\,, |x| \leq 1 \qquad \text{and} \qquad U^\mu(x) \simeq \frac{1}{|x|}\,, |x| \geq 1,$$

and so

$$E(\mu) = \int_{|x|\leq 1} U^\mu(x)\, d\mu\,(x) \simeq 1.$$

Here $A \simeq B$ means that there exists a constant $C \geq 1$, independent of the various parameters associated to the quantities A and B, such that $C^{-1}\,B \leq A \leq C\,B$.

We are now ready to use the notion of energy to present a sketch of the proof of the existence of an equilibrium measure. Given the conductor K, that is, given a compact subset K of $\mathbb{R}^3$, one tries to find an equilibrium distribution by considering a probability measure on K with minimal energy. In other words, one seeks a solution of the extremal problem

$$\inf\{\, E(\mu) : \|\mu\| = 1 \quad \text{and} \quad \operatorname{spt}\mu \subset K\},$$

where spt stands for "support," and tries to prove that it is an equilibrium distribution. A compactness argument in the unit ball of the space of finite measures in $\mathbb{R}^3$ easily shows that the infimum above is attained by some measure μ_e. A variational argument shows that the potential of μ_e is constant on K (and from here one can see that μ_e is unique with the property of attaining the infimum above). In fact this is so only when the boundary of K has some mild regularity, but we postpone the precise statement that holds in the general case until we have introduced the notion of capacity.

2.4. *Energy as a quadratic form in a Hilbert space*

One can easily see using Green's formula that, for some constant k,

$$\Delta\, U^\mu = -k\,\mu\,, \tag{1}$$

in the sense of distributions, Δ being the Laplace operator

$$\Delta = \sum_i \frac{\partial^2}{\partial x_i^2}\,.$$

Since (1) allows us to replace $d\mu(x)$ by $-k^{-1}\Delta\,\mu(x)dx$, an integration by parts shows that

$$\begin{aligned} E(\mu) &= \int U^\mu(x)\,d\mu(x) = -k^{-1}\int U^\mu(x)\,\Delta U^\mu(x)dx \\ &= -k^{-1}\sum_j \int U^\mu(x)\frac{\partial^2}{\partial x_j^2}\,U^\mu(x)\,dx \\ &= k^{-1}\sum_j \int \frac{\partial U^\mu}{\partial x_j}\,\frac{\partial U^\mu}{\partial x_j}\,dx \\ &= k^{-1}\int |\nabla U^\mu(x)|^2\,dx. \end{aligned}$$

This tells us that, modulo a fixed constant, the energy of a measure is the $L^2(\mathbb{R}^3)$ norm of the gradient of its Newtonian potential. Readers familiar with elementary properties of Hilbert spaces may now understand more clearly the uniqueness property of measures minimizing the energy functional. The preceding identity also hints at the prominent role played by Sobolev spaces in potential theory. Recall that the L^2-Sobolev space of order 1 is defined as

$$W^{1,2}(\mathbb{R}^3) = \{f \in L^2(\mathbb{R}^3) : \frac{\partial f}{\partial x_j} \in L^2(\mathbb{R}^3)\,,\ 1 \le j \le 3\}.$$

There is a far-reaching version of the notion of energy associated to an index $p \neq 2$ and the L^p-Sobolev space of order 1

$$W^{1,p}(\mathbb{R}^3) = \{f \in L^p(\mathbb{R}^3) : \frac{\partial f}{\partial x_j} \in L^p(\mathbb{R}^3)\,,\ 1 \le j \le 3\},$$

which gives rise to a vast "non-linear" potential theory (see [AH]), with interesting connections to some aspects of PDE.

2.5. *Capacity*

A central role in Electrostatics is played by "condensers." The simplest condenser consists of two plates: one of them connected to earth, and the second charged by a battery.

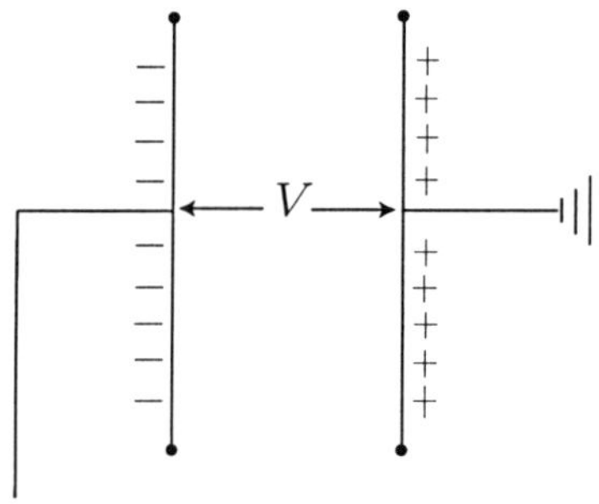

Figure 2

If one charges the right plate by an amount Q, then a difference of potential V appears between the plates. It turns out that experimental measurements show that the quotient $\frac{Q}{V}$ is constant and this constant is called the *capacity* of the condenser,

$$C = \frac{Q}{V}$$

Assume now that K is a conductor and that we take a charge distribution on K and let the charges move until they reach the equilibrium. Call μ the equilibrium distribution, so that $\mu(K)$ is the total charge. We know that the potential U^μ takes a constant value V on K. Define, following to Wiener, the **capacity** of K by

$$C(K) = \frac{\mu(K)}{V}.$$

One may imagine that the boundary of K and any sphere of large radius surrounding K are the plates of a condenser. Letting the radius of the sphere go to ∞, one gets an ideal condenser formed by the boundary of K and the point at ∞. Since the point at ∞ has zero potential ($U^\mu(\infty) = 0$), Wiener capacity may be understood as the capacity of this ideal condenser.

If you reach equilibrium from an initial unit charge distribution, then you get the equilibrium distribution μ_e of K. Its energy is precisely

$$E(\mu_e) = \int U^{\mu_e} \, d\,\mu_e = V.$$

Thus the constant value attained on K by the equilibrium distribution is exactly its energy. Since we learnt in the previous section that the equilibrium distribution minimizes energy, we obtain

$$C(K) = \frac{1}{V} = \frac{1}{\inf\ E(\mu)},$$

where the infimum is taken over the set of probability measures supported on K.

Sets of zero capacity play the role of negligible sets for potential theoretic questions, much in the same way as sets of zero measure are negligible in measure theory. Clearly K has zero capacity if and only if any (positive) measure supported on K has infinite energy. This happens for a point and also for a segment, although the segment case needs a bit of argument (see page 13). A sphere has positive capacity and a ball too, according to the examples discussed in subsection 1.4. In fact the capacity of a sphere is exactly its radius and the same happens for a ball. That the capacity of a ball is the same as the capacity of the sphere can be easily understood by the following heuristic argument : if you distribute the unit charge on a conductor ball, then to reach equilibrium the charges (which we assume to be positive) will go first to the boundary, pushed by the mutual forces acting among them by Coulomb's Law, and then will move in the sphere to reach equilibrium.

The fact that a ball and its boundary have the same capacity immediately implies that **capacity is not an additive set function**, that is, capacity is not a measure.

The complete solution of the problem of existence of an equilibrium distribution was found by Frostman in his thesis (1935). Frostman proved the following.

Theorem 2.1. *Given a compact set $K \subset \mathbb{R}^3$ there exists a unique probability measure μ on K such that U^μ is constant C-almost everywhere (that is, except for a set of zero capacity) on K.*

2.6. *Critical size of sets of zero capacity*

As we said before, a segment has zero capacity and a sphere has positive capacity. Now, a segment is a 1-dimensional object and a sphere is a 2-dimensional object. What can be said about sets with a fractional dimension between 1 and 2?.

The answer is as follows. If a set has (Hausdorff) dimension larger than 1, then it has positive capacity. If a set has finite 1-dimensional Hausdorff measure then it has zero capacity. Thus the critical dimension for capacity is exactly 1. Indeed, in dimension 1 one can find sets of vanishing capacity and also sets of positive capacity.

2.7. *Sub-additivity of capacity*

It turns out that capacity is a sub-additive set function in the sense that

$$C(K_1 \cup K_2) \leq C(K_1) + C(K_2),$$

for all compact sets K_1 and K_2.

This follows readily once one has an alternative description of capacity due to de La Vallée Poussin. Capacity is the maximal charge of a charge distribution with potential bounded by 1. In other words,

$$C(K) = \sup \{\mu(K) : \mu \geq 0, \ \mathrm{spt}\mu \subset K \text{ and } U^\mu(x) \leq 1, \ x \in \mathbb{C}\}.$$

Thus, in particular, a set K has positive capacity if and only if there is a positive measure supported on K with bounded potential.

To prove sub-additivity, take a positive measure μ supported on $K_1 \cup K_2$ such that $U^\mu(x) \leq 1$, $x \in K_1 \cup K_2$. Denote by μ_i the restriction of μ to K_i, $i = 1, 2$. Then obviously $U^{\mu_i}(x) \leq 1$, $x \in K_i$, $i = 1, 2$, and hence

$$\mu(K_1 \cup K_2) \leq \mu(K_1) + \mu(K_2) \leq C(K_1) + C(K_2).$$

Taking the supremum over such μ we get the sub-additivity inequality.

3. Analytic capacity

3.1. *Removable sets for bounded analytic functions*

A compact subset K of the plane is said to be **removable** (for bounded analytic functions) provided that for each open set $\Omega \supset K$ and any bounded analytic function f on $\Omega \setminus K$ one may extend f to an analytic function on Ω.

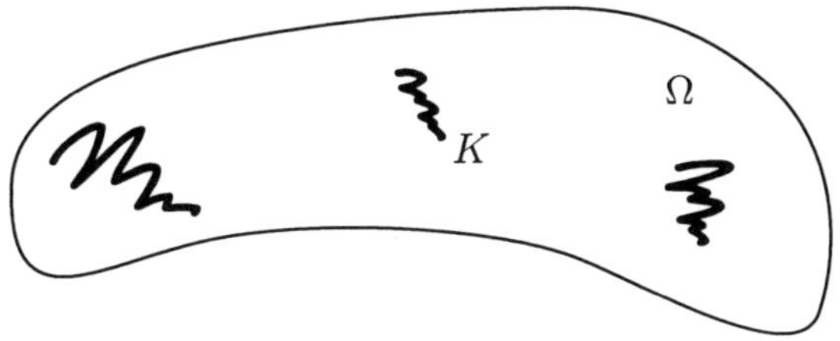

Figure 3

For example, a point is removable, according to a famous theorem of Riemann. A closed disc is not removable. Indeed, if the disc is centered at the origin it is clear that the function $f(z) = \frac{1}{z}$ is analytic and bounded on the complement of the disc, but it cannot be continued analytically to the whole plane.

It is not difficult to see that K is removable if and only if $H^\infty(\mathbb{C} \setminus K)$ (the set of bounded analytic analytic functions on $\mathbb{C} \setminus K$) is reduced to the constant functions. In other words, the test set Ω in the definition of removability can always be taken to be the whole plane.

The analytic capacity is a set function that quantifies the non-removability of a set. Since for a non-removable set there are non-constant bounded analytic functions on the complement of the set, it is natural to use derivatives to measure the non-removability of a set. The analytic capacity of a compact subset K of the plane is

$$\gamma(K) = \sup\{|f'(\infty)| : f \in H^\infty(\mathbb{C} \setminus K),\ \|f\|_\infty \leq 1 \text{ and } f(\infty) = 0\},$$

where $f'(\infty)$ is the derivative of f at ∞. A compactness argument provides a function attaining the above supremum, and the extremal function can be shown to be unique provided $f'(\infty) > 0$. As in the Wiener capacity case, it turns out that the critical dimension for analytic capacity is 1. This is simply due to the fact that the homogeneity of the Cauchy kernel $\frac{1}{z}$ is -1, exactly as in the case of the Newtonian kernel. One also has

$$\gamma(\bar{D}(a,r)) = \gamma(\partial D(a,r)) = r, \text{ for each } a \in \mathbb{C} \text{ and } r > 0.$$

Indeed, there is a striking formal analogy between C and γ, as shown by the expression

$$\gamma(K) = \sup\ |\langle T, 1\rangle|, \tag{2}$$

where the supremum is taken over those distributions T supported on K such that the distribution $\frac{1}{z} * T$ is a measurable bounded function on the plane and $||\frac{1}{z} * T||_\infty \leq 1$. This corresponds to the La Vallée Poussin expression for the Wiener capacity, with one important difference which will be discussed later on: positive measures are replaced by arbitrary distributions supported on K. The identity (2) is easy to prove. Given a bounded analytic function f on $\mathbb{C} \setminus K$ vanishing at ∞, one sets $T = \frac{1}{\pi}\bar{\partial} f$ so that $f = \frac{1}{z} * T$. In the opposite direction, given a distribution T with bounded Cauchy potential one simply sets $f = \frac{1}{z} * T$.

To stress the analogy between analytic and Wiener capacities, consider for compact subsets K of $\mathbb{R}^3$ the quantity

$$C^*(K) = \sup |\langle T, 1\rangle|, \tag{3}$$

where the supremum is taken over the class of distributions T supported on K, such that $\frac{1}{|x|} * T$ is a measurable bounded function on $\mathbb{R}^3$ and $||\frac{1}{|x|} * T||_\infty \leq 1$. Obviously $C(K) \leq C^*(K)$. We can easily show the following.

Proposition 3.1.

$$C^*(K) = C(K), \text{ for all compact } K \subset \mathbb{R}^3.$$

Proof. Consider a sequence of bounded open neighbourhoods Ω_n of K decreasing to K. We may assume, without loss of generality, that each Ω_n has a smooth boundary. Fix n and consider the equilibrium measure μ_e of the closure K_n of Ω_n. Then μ_e is a probability measure supported on K_n and its Newtonian potential takes a constant value V_n everywhere on K_n. Set $\mu = \frac{1}{V_n}\mu_e$, so that the Newtonian potential of μ is 1 everywhere on K_n. Take now a distribution T supported on K such that $||\frac{1}{|x|} * T||_\infty \leq 1$. We have

$$\langle T, 1\rangle = \langle T, \frac{1}{|x|} * \mu\rangle = \int (\frac{1}{|x|} * T)(x)\, d\mu(x),$$

and thus

$$|\langle T, 1\rangle| \leq ||\mu|| = \frac{1}{V_n} = C(K_n).$$

Letting n go to ∞ and then taking supremum on T we get

$$C^*(K) \leq C(K).$$ □

There is, however, a fundamental difference between analytic and Wiener capacities, which we now illustrate. As we said above, the Wiener capacity of a segment is 0. In terms of the de La Vallée Poussin expression for Wiener's capacity, this means that a segment does not support a non-zero positive measure with a bounded Newtonian potential. This can be shown as follows.

First of all we show that for any positive measure μ its Newtonian potential U^μ satisfies the identity

$$U^\mu(a) = \lim_{r\to 0} \frac{1}{|B(a,r)|} \int_{B(a,r)} U^\mu(x)\, dx,\ a \in \mathbb{R}^3, \tag{4}$$

where $|B(a,r)|$ stands for the volume of the ball of center a and radius r. Since $\frac{1}{|x|}$ is super-harmonic, so is $U^\mu = \frac{1}{|x|} * \mu$, and thus

$$U^\mu(a) \geq \frac{1}{\mid B(a,r) \mid} \int_{B(a,r)} U^\mu(x)\, dx,\ a \in \mathbb{R}^3,$$

which takes care of one inequality in (4). We obtain the other by observing that

$$\frac{1}{|a-y|} = \lim_{r\to 0} \frac{1}{|B(a,r)|} \int_{B(a,r)} \frac{1}{|x-y|}\, dx,\ a \neq y,$$

and applying Fatou's Lemma.

Assume now that μ is a positive measure supported on the segment $[0,1]$ (naturally identified with a subset of $\mathbb{R}^3$) and that its Newtonian potential is bounded almost everywhere with respect to Lebesgue measure in $\mathbb{R}^3$. Then (4) shows that U^μ is bounded everywhere in $\mathbb{R}^3$. If x is any point in $\mathbb{R}^3$, then

$$\frac{\mu B(x,r)}{r} \leq \int_{B(x,r)} \frac{d\mu(y)}{|x-y|}.$$

Since $\int \frac{d\mu(y)}{|x-y|} < \infty$, the preceding inequality shows that

$$\frac{\mu B(x,r)}{r} \to 0,\ \text{as } r \to 0.$$

Therefore μ is absolutely continuous with respect to Lebesgue measure on the segment and its density is identically 0.

On the contrary, the analytic capacity of a segment is positive. This can be shown readily by considering the conformal mapping of the complement of the segment in the Riemann sphere into the unit disc. This function is bounded analytic and non-constant on the complement of the segment.

In fact, one can construct explicit positive measures on the segment with bounded Cauchy potential. To do this first observe that the Cauchy potential of the Lebesgue measure on $[0,1]$ can be computed explicitly:

$$\int_0^1 \frac{dx}{z-x} = \log \frac{z}{z-1}, \; z \neq 0, \; z \neq 1.$$

Hence logarithmic singularities appear at 0 and 1. To get rid of them one can resort to a measure of the type $\mu = \varphi(x)dx$, where φ is a continuous function supported on the interval $[0,1]$, which is positive and linear on each of the intervals $(0, \frac{1}{2})$ and $(\frac{1}{2}, 1)$.

The reason behind the phenomenon described in the above paragraphs, which makes such a difference between analytic and Wiener capacities, is that the kernel $\frac{1}{|x|}$ is positive and thus only size matters in making the Newtonian potential of a positive measure bounded. Instead, since the Cauchy kernel $\frac{1}{z}$ is odd, cancellation phenomena may intervene in making the Cauchy potential of a positive measure bounded. This is precisely what happens in the case of the segment.

The conclusion is that in analyzing analytic capacity one must take into account cancellation issues. The Calderón-Zygmund theory of singular integrals is an example of how subtle this phenomena may be. The fact that the Cauchy kernel is complex is not as relevant as the fact that it is odd. Indeed, one can work with any of the real kernels $Re\frac{1}{z} = \frac{x}{|z|^2}, -Im\frac{1}{z} = \frac{y}{|z|^2}$, instead of $\frac{1}{z}$ and recover the theory of analytic capacity. In a word, analytic capacity is an object of a real variables nature and complex analysis may be practically ignored in studying it (see [3], [14] and [15]).

3.2. *Old problems on analytic capacity*

The definition of analytic capacity is due to Ahlfors (1947)(see [2]). Analytic capacity did not become an important object of study until Vitushkin showed in the middle sixties its fundamental role in understanding uniform approximation on compact subsets of the plane by rational functions with no poles on the set [16]. The problems raised by Vitushkin in his paper remained open for a long time. As the years passed without much progress, analytic capacity acquired an aura of mystery and was regarded as a difficult object to deal with. One of the reasons is that one is dealing with bounded analytic functions on arbitrary plane domains and bounded analytic functions are already difficult to understand in the unit disc. For the state of the art in 1994 see [13].

One of the problems raised by Vitushkin, known as **the Vitushkin conjecture**, inquired about a geometric description of the sets of zero analytic capacity among those having finite length (i.e., 1-dimensional Hausdorff measure). The conjecture was that a compact subset of the plane with finite positive length has zero analytic capacity if and only if its projections in almost all directions have zero length. These sets had been studied by Besicovitch in the thirties and are called Besicovitch irregular or purely unrectifiable. The terminology is explained by the fact that they are characterized by the property of having finite length and intersecting any rectifiable curve in a set of zero length. A breakthrough was made in (1996), where a special but significant case of Vitushkin's conjecture was proved [8]. The set was assumed to enjoy a homogeneity property called Ahlfors regularity, which consists in requiring that the length that one finds in the intersection of the set and a disc centered in the set is comparable to the radius of the disc (for discs of radius less than the diameter of the set). The proof made use of the Calderón-Zygmund theory of the Cauchy kernel with respect to the length measure on an Ahlfors regular set. The difficulty in extending the argument to the general case was related to the lack of the standard theory of singular integrals in non-homogeneous situations. This was overcome by G.David who proved the full conjecture in 1998 (see the survey [3]).

Another very well known old open question (this time not formulated by Vitushkin) was the characterization of the planar Cantor sets of zero analytic capacity (see [4]). This was achieved in (2003) (see [7]). In this paper one considered the problem of comparing analytic capacity with positive analytic capacity, which is defined by

$$\gamma_+(K) = \sup\{\mu(K) : \mu \geq 0\,,\ \mathrm{spt}\mu \subset K \text{ and } ||\frac{1}{z} * \mu||_\infty \leq 1\}.$$

Notice that trivially $\gamma_+(K) \leq \gamma(K)$, for each compact K. One was able to prove that

$$\gamma(K) \leq C_0\, \gamma_+(K), \tag{5}$$

for a family of approximations K of the Cantor set under consideration by finite unions of squares. The constant C_0 is independent of K.

In a remarkable article, Tolsa [11] proved some months later that (5) holds for all compact subsets of the plane . This showed that analytic capacity is semi-additive, because Tolsa himself had already proved in his thesis [10] that γ_+ is semi-additive .

More recent developments concern the bilipschitz invariance of analytic capacity, that is, the inequality

$$C^{-1}\,\gamma(K) \leq \gamma(\Phi(K)) \leq C\,\gamma(K),$$

where Φ is a mapping of the plane onto itself satisfying

$$C^{-1}\,|z-w| \le |\Phi(z)-\Phi(w)| \le C\,|z-w|.$$

Analytic capacity has been shown to be bilipschitz invariant in [12]. See also [5], where a partial result for planar Cantor sets was previously found.

3.3. *Some open problems*

There is still at least one apparently difficult open problem on analytic capacity, which consists in deciding whether a compact set of zero analytic capacity projects into zero length in almost all directions. There is a quantitative version of the problem, which we now formulate.

The **Favard length** of a compact set K in the plane (or its integral-geometric measure) is

$$F(K) = \frac{1}{\pi}\int_0^\pi |p_\theta(K)|\,d\theta,$$

where $p_\theta(K)$ stands for the projection of K into the straight line through the origin making an angle θ with the real axis, and $|p_\theta(K)|$ denotes its length.

It has been conjectured that there exists a number C_0 such that

$$F(K) \le C_0\,\gamma(K)$$

for all compact subsets K of $\mathbb{C}$.

Besides the above inequality, the challenging problem is now extending the semi-additivity inequality and the bilipschitz invariance property to higher dimensional real variables settings. The Cauchy kernel is replaced in $\mathbb{R}^n$ by the vector-valued Riesz kernels

$$R_\alpha(x) = \frac{x}{|x|^{1+\alpha}},\quad x\in\mathbb{R}^n,\ 0<\alpha<n.$$

The natural capacity associated to this kernel is

$$C_\alpha(K) = \sup\,|\langle T,1\rangle|,$$

where the supremum is taken over all distributions T supported on K such that the vector of distributions $R_\alpha * T$ is a measurable bounded vector valued function satisfying $||R_\alpha * T||_\infty \le 1$. The case $\alpha = n-1$ is especially interesting, because the well-known formula $(n>2)$

$$R_{n-1}(x) = c_n\,\nabla\Big(\frac{1}{|x|^{n-2}}\Big)(x),$$

establishes a relation between C_{n-1} and removable sets for Lipschitz harmonic functions, analogous to the relation between analytic capacity and removable sets for bounded analytic functions which has been discussed above.

Laura Prat begun a systematic study of the capacities C_α in her thesis [9]. A surprising result is that if $0 < \alpha < 1$, then a compact set of finite α-dimensional Hausdorff measure has zero C_α capacity. The same is true for a non-integer $1 < \alpha < n$ provided the set is Ahlfors regular of dimension α. The result should be true without this additional hypothesis.

In [6] it was shown that for $0 < \alpha < 1$, C_α is comparable to the capacity $C_{s,p}$ of non-linear potential theory, with smoothness index $s = \frac{2}{3}(n-\alpha)$ and L^p-index $p = \frac{3}{2}$. In particular, C_α is semi-additive and bilipschitz invariant for $0 < \alpha < 1$.

Volberg has shown in [17] that C_{n-1} is semi-additive in any dimension.

The two following two problems remain open.

Problem 1. Prove that C_α is semi-additive for all $0 < \alpha < n$. This should be feasible using the methods of [17].

Problem 2. Prove that for a non-integer α between 0 and n one has the inequality

$$C_\alpha^{-1}\, C_{\frac{2}{3}(n-\alpha),\frac{3}{2}}(K) \leq C_\alpha(K) \leq C_\alpha\, C_{\frac{2}{3}(n-\alpha),\frac{3}{2}}(K),$$

for all compact subsets K of $\mathbb{R}^n$, C_α being a constant independent of K. This seems to be a much more challenging problem, because Riesz kernels R_α with $\alpha > 1$ do not have the convenient subtle positivity properties that one finds in the case $0 < \alpha \leq 1$. See [6] and [9].

Acknowledgments

The author is grateful for the generous support provided by the grants MTM-2004-00519 (DGCYT), Acción integrada HF2004-0208 and 2005 SGR 00774 (Generalitat de Catalunya). Thanks are also due to my colleagues L. Narici and A.G. O'Farrell for correcting English in the first draft of the manuscript.

References

1. D.R. Adams and L.I. Hedberg, *Function Spaces and Potential Theory*, Grundlehren der Mathematischen Wissenschaften, 314. Springer-Verlag, Berlin, 1996.

2. L. Ahlfors, *Bounded analytic functions*, Duke Math. J. **14** (1947),1–11.
3. G. David, *Analytic capacity, Calderón-Zygmund operators, and rectifiability*, Publ. Mat. **43**(1999), 3–25.
4. J. Garnett, *Analytic capcity and measure*, Lec. Notes in Math. **297**, Springer, Berlin-New York, 1972.
5. J. Garnett and J. Verdera, *Analytic capacity, bilipschitz maps and Cantor sets*, Math. Res. Lett. **10**(2003), 515–522.
6. J.Mateu, L.Prat and J.Verdera, *The capacity associated to signed Riesz kernels, and Wollf's potentials*, J. Reine Angew Math. **578** (2005), 201-223.
7. J. Mateu, X. Tolsa and J. Verdera, *The planar Cantor sets of zero analytic capacity and the local $T(b)$-theorem*, J. Amer. Math. Soc. **16**(2003), 19–28.
8. P. Mattila, M.S. Melnikov and J. Verdera, *The Cauchy integral, analytic capacity, and uniform rectifiability*, Annals of Math. **144** (1996), 127-136.
9. L. Prat, *Potential theory of signed Riesz kernels: capacity and Hausdorff measure*, Intern. Math. Res. Not. **19** (2004), 937–981.
10. X. Tolsa, *L^2-boundedness of the Cauchy integral operator for continuous measures*, Duke Math. J. **98(2)**(1999), 269–304.
11. X. Tolsa, *Painleve's problem and the semi-additivity of analytic capacity*, Acta Math. **190** (2003), 105–149.
12. X. Tolsa, *Bilipschitz maps, analytic capacity and the Cauchy integral*, Annals of Math. (to appear).
13. J. Verdera, *Removability, capacity and approximation,* in *Complex potential theory*, NATO Adv. Sci. Inst. Ser. C Math. Phys. Sci., **439** (1994), 419–473.
14. J. Verdera, *L^2 boundedness of the Cauchy integral and Menger curvature*, Contemp. Math. **277** (2001), 139–158.
15. J. Verdera *Ensembles effaçables, ensembles invisibles et le problème du voyageur de commerce, ou Comment l'analyse réelle aide l'analyse complexe*, Gaz. Math. **101** (2004), 21–49.
16. A. G. Vitushkin, *The analytic capacity of sets in problems of approximation theory*, Russian Math. Surveys **22**(1967), 139–200.
17. A. Volberg, *Calderón-Zygmund capacities and operators on nonhomogeneous spaces*, CBMS Reg. Conf. Ser. Math. **100**, American Mathematical Society, Providence, RI, 2003.
18. J.Wermer, *Potential theory*, Lecture Notes in Mathematics, **408**, Springer-Verlag, Berlin-New York, 1974.

BEST APPROXIMATIONS ON SMALL REGIONS. A GENERAL APPROACH

FELIPE ZÓ

IMASL-CONICET and Universidad Nacional de San Luis.
5700-San Luis. Argentina.
E-mail: fzo@unsl.edu.ar

HÉCTOR H. CUENYA

Universidad Nacional de Río Cuarto.
5800-Río Cuarto. Argentina.
E-mail: hcuenya@exa.unrc.edu.ar

For a family of monotone function seminorms acting on Lebesgue measurable functions $F : B \subset I\!R^n \to I\!R^k$, $B = \{x \in I\!R^n : |x| \leq 1\}$, we consider the best approximation function $P_\varepsilon \in A$, where A is some subspace of polynomials. The asymptotic behavior of a normalized error function as well as the limit of the net $\{P_\varepsilon\}$ as $\varepsilon \to 0$ are studied. This vector valued approach extends and unifies some classical problems in best local approximation theory.

Keywords: Best local approximation, function norms, weak convergence.
AMS Classification: 41A65.

1. Introduction

The first association between the best approximation polynomials and the Taylor polynomial is given in [16], where it is proved the following. "If I is an interval (variable) of the axis of reals, and if $f(z)$ is real for real z and analytic at the origin, then as I approaches the origin the function $P_{m,I}(z)$ approaches the function $T_m(f)(z)$, uniformly for z on any limited point set". The function $T_m(f)$ is the Taylor polynomial of f at the origin, and $P_{m,I}$ is the polynomial of degree at most m which minimizes

$$\|f - P\|_{L^\infty(I)}, \; P \in \pi^m.$$

In the same paper J. Walsh observed that the actual analyticity of the function $f(z)$ at the origin need not be assumed; the existence and continuity of certain derivatives is in fact sufficient. There undoubtedly exist other ways in which (perhaps the most general of) the functions $T_m(f)$ can be considered as limiting cases of the functions $P_{m,I}$.

At the present it is known that the classical concept of Peano's derivative [18] it suffices to assure the Walsh result, and the norm L^∞ can be replaced by L^p or Orlicz norms, see [7], [17], [3], [12], [9] and more recently in [4]. It is a rather obvious fact that the smoothness required for the function f should be accord with the norm used in the best approximation problem. Thus if we are considering a best approximation obtained with an L^p norm the concept of derivative at a point in the L^p sense, extensively studied in [8], should be used in order to assume less smoothness on the function f.

In section 3 we define a general concept of Peano's derivative within a frame of a family of function seminorms $\|F\|_\varepsilon$, $0 \leq \varepsilon \leq 1$, acting on Lebesgue measurable functions $F : B \subset \mathbb{R}^n \to \mathbb{R}^k$, here $B = \{x \in \mathbb{R}^n : |x| \leq 1\}$, and $|\cdot|$ denotes the euclidean norm on $\mathbb{R}^n$. This concept of derivative appeared in [8], [14], and [11] for different L^p norms; however a somewhat more general concept of smoothness is used in [4].

The idea of shrinking sets $\{V_\varepsilon\}$ to a single point and to further study the behavior of a best approximation net $\{P_\varepsilon(f)\}$ obtained by minimizing a certain norm on V_ε using a fixed approximation class of functions has been extended to several points. As an example we state the next result which is a particular case of a more general situation appeared in [2]. For $-1 < x_1 < ... < x_k < 1$ let V_ε be the union of the intervals centered at the points x_i and length 2ε. Given $1 < p < \infty$, let p_ε be the polynomial in π^l, the set of algebraic polynomials of degree at most l, which minimizes $\|f-p\|_{L^p(V_\varepsilon)}$, and p in π^l. For $(m+1)k-1 < l \leq (m+2)k-1$, let $f \in C^{m+1}$ near each x_i; then $p_0 = \lim_{\varepsilon\to 0} p_\varepsilon$ exists. The polynomial p_0 is usually called the best local approximation or a multipoint best local approximation. A different proof of the above result appeared in [12]. Moreover the best local approximation p_0 is characterized just in term of the derivatives $f^{(j)}(x_i)$, for $0 \leq j \leq m+1$, $1 \leq i \leq k$, as follows. Let π_0^l be defined by $\{p \in \pi^l : p^{(j)}(x_i) = 0, 0 \leq j \leq m, 1 \leq i \leq k\}$. The polynomial p_0 is the sum $h + q$ with $h \in \pi^{(m+1)k-1}$ and $q \in \pi_0^l$. The first one is determined by the interpolating conditions $h^{(j)}(x_i) = f^{(j)}(x_i)$, for $0 \leq j \leq m$, $1 \leq i \leq k$, and the second one is the solution to the problem

$q \in \pi_0^l$, and $(q^{(m+1)}(x_1), ..., q^{(m+1)}(x_k))$ is the minimun of

$$|f^{(m+1)}(x_1) - y_1|^p + ... + |f^{(m+1)}(x_k) - y_k|^p,$$

where $(y_1, ..., y_k)$ is in the subspace defined by

$$\{(p^{(m+1)}(x_1), ..., p^{(m+1)}(x_k)) : p \in \pi_0^l\} \subseteq \mathbb{R}^k.$$

We have this result in a slight general version at the very end of the paper.

We shall deal with best approximations $\{P_\varepsilon(F)\}$ for a vector valued function F of several variables. The approximation class will be a finite subspace $\mathcal{A}$ of vector valued algebraic polynomials of several variables with some conditions which are introduced in section 4. The case of multipoint best local approximation quoted above is covered by the class of polynomials $\mathcal{A}(l,k)$ of section 4, and new $\mathcal{A}$ are showed.

The best approximations $\{P_\varepsilon(F)\}$ are taken with respect to a general monotone family of seminorms introduced in section 2 which strictly covers the norms used in dealing with best local approximation problems. Classical norms computed on regular regions $\{V_\varepsilon\}$ is a particular case of our set up.

The technical matter of this paper is to obtain what we call the asymptotic behavior of the error function, see Theorems 4.2 and 4.5. The first result is in [13] and see also [14], [2] and [4]. In our context two different types of approximation classes $\mathcal{A}$ appear clearly. In Theorem 4.2 we have a neat expression for the asymptotic behavior of the error function which is not so nice in Theorem 4.5, except for particular cases of seminorms used in a problem.

We believe that this unified approach to best local approximation problems has the next two advantages over the other views. First, it separates the approximation problem of the technical conditions of the family of seminorms. Secondly, to study vector valued functions F with family of approximations $\mathcal{A}$ brings new possibilities of research.

2. The norm set up

We will work with a family of function seminorms $\|F\|_\varepsilon$, $0 \le \varepsilon \le 1$, acting on Lebesgue measurable functions $F : B \subset I\!R^n \to I\!R^k$, where $B = \{x \in I\!R^n : |x| \le 1\}$, and $|\cdot|$ denotes the euclidean norm on $I\!R^n$. Therefore it may happen $\|F\|_\varepsilon = 0$ for a non zero function F and $\varepsilon > 0$; however in most of the applications the "limit" function seminorm $\|\cdot\|_0$ will be a norm.

We assume the following properties for the family of function seminorms $\|F\|_\varepsilon$, $0 \le \varepsilon \le 1$.

(1). For $F = (f_1, ... f_k)$, and $G = (g_1, ..., g_k)$, we have $\|F\|_\varepsilon \le \|G\|_\varepsilon$ for every $\varepsilon > 0$, provided $|f_i(x)| \le |g_i(x)|$, $i = 1, ..., k$, and $x \in B$.

(2). If 1 is the function $F(x) = (1, ..., 1)$, we have $\|1\|_\varepsilon < \infty$, for all $\varepsilon \ge 0$.

(3). For every $F \in C_k(B)$, we have $\|F\|_\varepsilon \to \|F\|_0$, as $\varepsilon \to 0$, where

$C_k(B)$ is the set of continuous functions $F : B \subset I\!R^n \to R^k$. Moreover $\|F\|_0$ is a norm on $C_k(B)$.

¿From now on, if we do not specify the contrary, the statements will be valid for an abstract family of seminorms $\|F\|_\varepsilon$, $0 \le \varepsilon \le 1$, fulfilling conditions (1)-(3).

In order to give examples of norms $\|F\|_\varepsilon$, $0 \le \varepsilon \le 1$, with the properties (1)-(3) we recall a definition of convergence of measures early given in [10]. See also [1] for the notion of weak convergence of measures in general.

Definition 2.1. Let μ_ε , $0 \le \varepsilon \le 1$, be a family of probability measures on B. We say that the measures μ_ε converge weakly in the proper sense to the measure μ_0 if we have

$$\int_B f(x)\, d\mu_\varepsilon(x) \to \int_B f(x)\, d\mu_0(x), \quad f \in C_1(B),$$

and $\mu_0(B') > 0$ for any ball $B' \subseteq B$.

The assumption on the measure μ_0 implies that

$$\|F\|_\varepsilon = \|F\|_{L^p(\mu_\varepsilon)} = (\int_B \|F\|^p d\mu_\varepsilon)^{1/p},$$

is actually a norm on $C_k(B)$ for $\varepsilon = 0$ and $1 \le p < \infty$, where $\|\cdot\|$ stands for any monotone norm on $I\!R^k$. A norm (seminorm) fulfilling a property like (1) is called a monotone norm. We use a monotone norm on $I\!R^k$ to assure property (1) for the family of seminorms $\|F\|_\varepsilon$, $0 \le \varepsilon \le 1$. It is worthy to note we will not need this property on $\|\cdot\|$ in proving the convergence results in the coming Propositions 2.1, 2.2 and 2.5.

Let F be in $C_k(B)$; it is readily seen, by using the definition of weak convergence of measures, that there exists $\varepsilon_0 = \varepsilon_0(F) > 0$ such that if $\|F\|_\varepsilon = \|F\|_{L^p(\mu_\varepsilon)} = 0$, for some $0 < \varepsilon \le \varepsilon_0$ then $F = 0$. The following convergence result is also clear.

Proposition 2.1. *Let μ_ε , $0 \le \varepsilon \le 1$, be a family of probability measures on B which converges weakly in the proper sense to the measure μ_0. Then $\|F\|_\varepsilon = \|F\|_{L^p(\mu_\varepsilon)}$ converges as $\varepsilon \to 0$ to the norm $\|F\|_0 = \|F\|_{L^p(\mu_0)}$ if $F \in C_k(B)$.*

An example of measures μ_ε is the one given in [14] by

$$\mu_\varepsilon(E) = \int_E \varepsilon^n w(\varepsilon t) W^{-1}(\varepsilon) dt,$$

where $W(\varepsilon) = \int_{|t|\leq\varepsilon} w(t)dt$, and $\varepsilon t = (\varepsilon t_1, \varepsilon t_2, ..., \varepsilon t_n)$. The following condition was assumed on the weight function w:

$$W(\varepsilon) = A\varepsilon^{\beta+n}(1 + o(1)), \quad \text{as} \quad \varepsilon \to 0, \quad A > 0, \quad \beta + n > 0.$$

It is proved in [11], for radial weight functions w, that these measures μ_ε converge weakly to the measure

$$\mu_0(E) = \omega_n^{-1}(\beta + n) \int_E |t|^\beta dt,$$

where ω_n is the surface measure of the unit sphere in $I\!R^n$. Other examples of measures are found in [11].

For the next examples we denote by μ_ε a family of abstract probability measures which converge weakly in the proper sense to a measure μ_0.

Proposition 2.2. *Let φ be a convex function such that $\varphi(0) = 0$, $\varphi(x) > 0$ if $x > 0$. For any measurable function $F : B \subset I\!R^n \to I\!R^k$, set*

$$\|F\|_\varepsilon = \inf\{\lambda > 0 : \int_B \varphi\left(\frac{\|F(t)\|}{\lambda}\right) d\mu_\varepsilon(t) \leq 1\},$$

where $\|\cdot\|$ is any norm on $I\!R^k$. Then $\|F\|_\varepsilon$ converges to $\|F\|_0$, for any $F \in C_k(B)$.

Proof. Let F be in $C_k(B)$ a nonzero function; then there exist constants $\varepsilon_0 > 0$ and c, both depending on F, such that $1 \leq c\|F\|_\varepsilon \leq c^2$ for all $0 \leq \varepsilon \leq \varepsilon_0$. Let us see first that $\|F\|_\varepsilon$ is bounded away from zero. If this were not so, for any $\delta > 0$ we can find a sequence $\varepsilon_k \searrow 0$ such that $\|F\|_{\varepsilon_k} \leq \delta$. Then

$$1 \geq \int_B \varphi\left(\frac{\|F(t)\|}{\|F\|_{\varepsilon_k} + \delta}\right) d\mu_{\varepsilon_k}(t) \geq \int_B \varphi\left(\frac{\|F(t)\|}{2\delta}\right) d\mu_{\varepsilon_k}(t).$$

Therefore $\int_B \varphi(\frac{\|F(t)\|}{2\delta})d\mu_o(t) \leq 1$, then $\|F\|_0 \leq 2\delta$ for every $\delta > 0$, which is a contradiction. If $\|F\|_\varepsilon$ were not bounded from above we can find a sequence $\varepsilon_k \searrow 0$ such that $\|F\|_{\varepsilon_k} \geq k$. Then for every k,

$$1 = \int_B \varphi\left(\frac{\|F(t)\|}{\|F\|_{\varepsilon_k}}\right) d\mu_{\varepsilon_k}(t) \leq \int_B \varphi\left(\frac{\|F(t)\|}{k}\right) d\mu_{\varepsilon_k}(t)$$

$$\leq \frac{1}{k} \int_B \varphi(\|F(t)\|)d\mu_{\varepsilon_k}(t) \leq \frac{C}{k},$$

where $C = \max_{t\in B}\varphi(\|F(t)\|)$, which is a contradiction. Let $\varepsilon_j \to 0$ be such that $\|F\|_{\varepsilon_j} \to \lambda$. Clearly $\lambda > 0$ and

$$1 = \int_B \varphi\left(\frac{\|F(t)\|}{\|F\|_{\varepsilon_j}}\right) d\mu_{\varepsilon_j}(t) \to \int_B \varphi\left(\frac{\|F(t)\|}{\lambda}\right) d\mu_0(t),$$

as $j \to \infty$. But the equality $\int_B \varphi(\frac{\|F(t)\|}{\lambda}) d\mu_0(t) = 1$ always implies that $\lambda = \|F\|_0$. □

The following example will be of interest later on.

Proposition 2.3. *Let φ be a convex function such that $\varphi(0) = 0,\ \varphi(x) > 0$ if $x > 0$. Then $\|F\|_\varepsilon$ converges to $\|F\|_0$, for any $F \in C_k(B)$, where*

$$\|F\|_\varepsilon = \inf\{\lambda > 0 : \int_B \sum_{i=1}^k \varphi\left(\frac{|f_i(t)|}{\lambda}\right) d\mu_\varepsilon(t) \le 1\},$$

and $F(t) = (f_1(t), ..., f_k(t))$.

The proof is analogous to that of Proposition 2.2.

The next abstract example will be used at the end of section 5.

Proposition 2.4. *Let $\|\cdot\|_{i,\varepsilon},\ 1 \le i \le k$, and $0 \le \varepsilon < 1$, be a family of seminorms fulfilling (1), (2) and (3). Set for $F(t) = (f_1(t), ..., f_2(t))$ the seminorms $\|F\|_\varepsilon = \|(\|f_1\|_{1,\varepsilon}, ..., \|f_k\|_{k,\varepsilon})\|$, where $\|\cdot\|$ is a monotone norm on $\mathbb{R}^k$. Then $\|F\|_\varepsilon,\ 0 \le \varepsilon \le 1$, has the properties (1), (2) and (3).*

For the following Orlicz norm,

$$\|F\|_\varepsilon = \inf_{s>0} \frac{1}{s}(1 + \int_B \varphi(s\|F(t)\|)\ d\mu_\varepsilon(t)),$$

we have a similar result to Proposition 2.3, but more conditions on the convex function φ are needed.

Proposition 2.5. *Let φ be a convex function such that $\varphi(0) = 0,\ \varphi(x) > 0$ if $x > 0$, and $\frac{\varphi(x)}{x} \to \infty$ as $x \to \infty$. Then $\|F\|_\varepsilon \to \|F\|_0, F \in C_k(B), \varepsilon \to 0$.*

Proof. There exists $s_\varepsilon > 0$ such that

$$\|F\|_\varepsilon = \frac{1}{s_\varepsilon}(1 + \int_B \varphi(s_\varepsilon(\|F(t)\|)\ d\mu_\varepsilon(t)).$$

The norms $\|F\|_\varepsilon$ are bounded by $1 + \max_{x\in B}\varphi(\|F(x)\|)$ for $0 \le \varepsilon \le 1$. Then the net (s_ε) is bounded away from zero for small ε since $\|F\|_\varepsilon \ge \frac{1}{s_\varepsilon}$.

We will see that (s_ε) is bounded from above for small ε. In fact, let B' be a ball in B such that $m = \min_{B'} \|F(t)\| > 0$; then

$$\|F\|_\varepsilon \geq \int_B \frac{\varphi(s_\varepsilon \|F(t)\|)}{s_\varepsilon} d\mu_\varepsilon(t) \geq \frac{\varphi(s_\varepsilon m)}{s_\varepsilon} \mu_\varepsilon(B').$$

As $\mu_0(B') = \mu_\varepsilon(B') + o(1)$ as $\varepsilon \to 0$, we have

$$\mu_0(B') \leq \frac{s_\varepsilon}{\varphi(s_\varepsilon m)}(1 + \max_{x \in B} \varphi(\|F(x)\|)) + o(1).$$

If we suppose (s_ε) has not an upper bound we would have $\mu_0(B') = 0$ since $\frac{\varphi(x)}{x} \to \infty$.

Now assume $s_{\varepsilon_l} \to s_0$; then $s_0 > 0$ and

$$\|F\|_0 \leq \frac{1}{s_0}\left(1 + \int_B \varphi(s_0 \|F(t)\|)\ d\mu_0(t)\right)$$

$$\leq \frac{1}{s_{\varepsilon_l}}\left(1 + \int_B \varphi(s_{\varepsilon_l} \|F(t)\|)\ d\mu_{\varepsilon_l}(t)\right) + o(1).$$

On the other hand, given $s_1 > 0$ such that

$$\|F\|_0 = \frac{1}{s_1}\left(1 + \int_B \varphi(s_1 \|F(t)\|)\ d\mu_0(t)\right),$$

we have

$$\|F\|_{\varepsilon_l} \leq \frac{1}{s_1}\left(1 + \int_B \varphi(s_1 \|F(t)\|)\ d\mu_{\varepsilon_l}(t)\right) =$$

$$\frac{1}{s_1}\left(1 + \int_B \varphi(s_1 \|F(t)\|)\ d\mu_0(t)\right) + o(1) = \|F\|_0 + o(1).$$

Therefore $\|F\|_\varepsilon \to \|F\|_0$, as $\varepsilon \to 0$. □

3. The Taylor polynomial and the limit of best approximation polynomials

We denote by π^m the set of all algebraic polynomials in n-variables of degree at most m, and Π_k^m the set $\{P = (p_1, ..., p_k) : p_i \in \pi^m\}$.

Given a function $F : B \subset \mathbb{R}^n \to \mathbb{R}^k$, and a family of seminorms $\|F\|_\varepsilon$, $0 \leq \varepsilon \leq 1$, as in section 2, we introduce a general version of the "Peano's definition" of the Taylor polynomial. We will use the notation $F^\varepsilon(x) = F(\varepsilon x)$.

Definition 3.1. A function $F : B \subset I\!R^n \to I\!R^k$, has a Taylor Polynomial of degree m if there exists $T_m = T_m(F) \in \Pi_k^m$ such that

$$\|F^\varepsilon - T_m^\varepsilon\|_\varepsilon = o(\varepsilon^m), \quad \varepsilon \to 0.$$

To prove the uniqueness of the Taylor polynomial we need the next result which is a consequence of a usual compactness argument which is included for the sake of completeness.

Proposition 3.1. *There exist $C = C(m,k)$ and $0 < \varepsilon(m,k)$ such that for every $0 < \varepsilon \leq \varepsilon(m,k)$,*

$$C^{-1}\|P\|_0 \leq \|P\|_\varepsilon \leq C\|P\|_0,$$

for every $P \in \Pi_k^m$.

Proof. If the right hand side of the above inequality were not true, we can select sequences $\varepsilon_l \searrow 0$, and $P_l \in \Pi_k^m$ with $\|P_l\|_0 = 1$ and $\|P_l\|_{\varepsilon_l} \geq l$. We may assume, by taking a new subsequence if it is necessary, that P_l converges uniformly on B to a polynomial $P_0 \in \Pi_k^m$. But

$$|\|P_l\|_{\varepsilon_l} - 1| \leq \|P_l - P_0\|_{\varepsilon_l} + |\|P_0\|_{\varepsilon_l} - \|P_0\|_0|.$$

Note that $\|P_l - P_0\|_{\varepsilon_l} \to 0$ as $l \to \infty$ since we are dealing with monotone seminorms. By the property (3) of the family of seminorms, $\|P_0\|_{\varepsilon_l} \to \|P_0\|_0 = 1$. Thus $\|P_l\|_{\varepsilon_l} \to 1$, which is a contradiction.

The procedure to prove the remaining inequality is similar to the previous one. □

We obtain the following useful inequalities by using Proposition 3.1 for the polynomial P^ε with the seminorm $\|\partial^\alpha P(0)\|$.

Proposition 3.2. *There exists a constant $C > 0$, depending only on m, k, n and the family of norms $\|\cdot\|_\varepsilon$ such that for any $P = (p_1, ..., p_k) \in \Pi_k^m$ there holds*

$$|\partial^\alpha p_i(0)| \leq C\varepsilon^{-|\alpha|}\|P^\varepsilon\|_\varepsilon,$$

for any $0 \leq \alpha \leq m$, and $i = 1, ..., k$.

As a direct consequence of Proposition 3.2 we have

Proposition 3.3. *The polynomial $T_m = T_m(F) \in \Pi_k^m$ in definition 3.1, if there exists, is unique.*

Recall that a real function f defined on $(-1,1)$ has a Peano's derivative α_m of order m at 0 if there exist constants $\alpha_0, \alpha_1, ..., \alpha_m$ such that for small $|t|$,

$$f(t) = \alpha_0 + \alpha_1 t + ... + \frac{\alpha_{m-1}}{(m-1)!} t^{m-1} + \frac{\alpha_m + \varepsilon_m}{m!} t^m,$$

where $\varepsilon_m \to 0$ with t. The above concept has a direct extension to several variables. See [18] for classical references on Peano's derivatives.

Proposition 3.4. *Let f_i, $i = 1, ..., k$ be real valued functions on $[-1,1]$ such that each of them has a m-th Peano's derivative at 0. Then $F = (f_1, ..., f_k)$ has a Taylor polynomial $T_m = T_m(F)$ of degree m, for each family of seminorns $\|F\|_\varepsilon$, $0 \leq \varepsilon \leq 1$.*

Proposition 3.5. *If the function F has the Taylor polynomial of degree m, $T_m(x) = \sum_{0 \leq |\alpha| \leq m} A_\alpha x^\alpha$, then the Taylor polynomial of degree $l \leq m$ is given by $T_l(x) = \sum_{0 \leq |\alpha| \leq l} A_\alpha x^\alpha$. We set $\partial^\alpha F(0)$ for the vector $\alpha! A_\alpha$.*

Proof. In fact,

$$\|F^\varepsilon - T^\varepsilon_{m-1}\|_\varepsilon \leq \|F^\varepsilon - T^\varepsilon_m\|_\varepsilon + \|T^\varepsilon_m - T^\varepsilon_{m-1}\|_\varepsilon$$

$$\leq o(\varepsilon^m) + \| \sum_{|\alpha|=m} A_\alpha x^\alpha \|_\varepsilon = o(\varepsilon^m) + O(\varepsilon^m) = o(\varepsilon^{m-1}),\ (\varepsilon \to 0).$$ □

Notation 3.1. We write $F \in t^m$ if the function F has the Taylor polynomial of order m.

The class t^m was extensively studied by A.P. Calderón and A. Zygmund in [8] for the case $\|f\|_\varepsilon = \|f\|_{L^p(B)}$; here the homogeneous dilations $\varepsilon x = (\varepsilon x_1, ..., \varepsilon x_n)$ were also used in the definition of the Taylor polynomial. In the case of non homogeneous dilations $\varepsilon_\alpha(x) = (\varepsilon^{\alpha_1} x_1, ..., \varepsilon^{\alpha_n} x_n)$ it is also possible, and useful, to consider the corresponding Peano's version of the Taylor polynomials, see [15] and [5]. The use of the class t^m in approximation problems appears in [14] and further in [11] and [12]; however this concept was implicitly used by other authors in best local approximation problems.

For $x_0 \in B$, let $G(s) = F(x_0 + s)$, and $T_m = T_m(G)$. Then $T_m(x - x_0)$ is the usual Taylor polynomial of F at x_0.

Notation 3.2. For $-1 < x_1 < ... < x_k < 1$ and $f : [-2,2] \to I\!R$ set $F(s) = Lf(s) = (f(x_1 + s), ..., f(x_k + s))$.

Now, let $\|F\|_\varepsilon$, $0 \le \varepsilon \le 1$, be a family of seminorms for functions $F : B \subset I\!R \to I\!R^k$. Then we have

Proposition 3.6. *There exists* $h \in \pi^{(m+1)k-1}$ *such that* $\|(L(f) - L(h))^\varepsilon\|_\varepsilon = o(\varepsilon^m)$, *as* $\varepsilon \to 0$, *if and only if the function* $F = L(f)$ *is in* t^m, *i.e.,* $L(f)$ *has a Taylor polynomial of order* m. *Moreover, if there exists* h, *it is unique and it is called the Hermite polynomial which interpolates the data* $f^{(j)}(x_i)$, *and* $0 \le j \le m$, $1 \le i \le k$.

Proof. For $P = (p_1, ..., p_k) \in \Pi_k^m$ set $h = I(P)$ for the unique polynomial $h \in \pi^{(m+1)k-1}$ such that $h^{(j)}(x_i) = p_i^{(j)}(0)$ for $0 \le j \le m$, $\quad i = 1, ..., k$. The function $I : \Pi_k^m \longrightarrow \pi^{(m+1)k-1}$ is an isomorphism and $\|(P - L(h))^\varepsilon\|_\varepsilon = O(\varepsilon^{m+1})$, as $\varepsilon \to 0$, for any family of seminorms $\|F\|_\varepsilon$, $0 \le \varepsilon \le 1$. Let $h \in \pi^{(m+1)k-1}$ be such that $\|(L(f) - L(h))^\varepsilon\|_\varepsilon = o(\varepsilon^m)$, and consider $P \in \Pi_k^m$ with $I(P) = h$. Then $\|(P - L(f))^\varepsilon\|_\varepsilon \le \|(L(f) - L(h))^\varepsilon\|_\varepsilon + O(\varepsilon^{(m+1)}) = o(\varepsilon^m)$, i.e., Lf has the Taylor polynomial P. The rest of the proof is completed in a similar way. $\square$

We observe that the notation t^m does not show the points x_i of the examples.

Let $\|F\|_\varepsilon$, $0 \le \varepsilon \le 1$, be a family of seminorms and $F : B \subset I\!R^n \to R^k$, be a fixed measurable function such that $\|F\|_\varepsilon$ and $\|F^\varepsilon\|_\varepsilon$ are finite for all ε.

Definition 3.2. Set $\|F\|_\varepsilon^* = \|F^\varepsilon\|_\varepsilon$, and $P_\varepsilon = P_\varepsilon(F)$ for any polynomial in Π_k^m which minimizes $\|F - P\|_\varepsilon^*$, $P \in \Pi_k^m$.

Although the best approximation polynomial $P_\varepsilon(F)$ is not unique in general, through this paper the notation $P_\varepsilon(F)$ does not mean a set of best approximation polynomials but any arbitrarily chosen polynomial in this set. We have the existence of $P_\varepsilon(F)$, at least for all small ε, by Proposition 3.1.

The next statement has its origin in [16] using the L^∞ norm, and since then similar versions in L^p in one and several real variables appeared. Results dealing with weighted Luxemburg norms appeared recently in [4], in particular see Theorem 3.3 which is generalized by Theorem 3.1.

Theorem 3.1. *If* $F \in t^m$, *then* $P_\varepsilon \to T_m(F)$ *as* $\varepsilon \to 0$.

Proof. In fact, $\|P^\varepsilon - T_m^\varepsilon(F)\|_\varepsilon \le 2\|F^\varepsilon - T_m^\varepsilon(F)\|_\varepsilon = o(\varepsilon^m)$, and by Proposition 3.2 it follows $|P_{\varepsilon,i}^{(\beta)}(0) - T_{m,i}^{(\beta)}(F)(0)| \le \varepsilon^{-|\beta|} C \|P_\varepsilon^\varepsilon - T_m^\varepsilon(F)\|_\varepsilon$, and $|\beta| \le m$. $\square$

Chui called best local approximation of F the limit of $P_\varepsilon(F)$ as $\varepsilon \to 0$ (see [6]).

For the case $B = [-1,1]$ and $-1 < x_1 < ... < x_k < 1$ and $F(s) = Lf(s) = (f(x_1+s),...,f(x_k+s)) = (f_1(s),...,f_k(s))$, we consider, as an example, the following classic L^p norm, although similar conclusions hold for some other norms.

$$\|F\|_\varepsilon = \|F\| = \left(\sum_{i=1}^{k} \int_B |f_i(x)|^p \; dx\right)^{\frac{1}{p}} =$$

$$\left(\sum_{i=1}^{k} \int_B |f(x_i+x)|^p \; dx\right)^{\frac{1}{p}} = \left(\sum_{i=1}^{k} \int_{B(x_i)} |f(x)|^p \; dx\right)^{\frac{1}{p}},$$

where dx is the Lebesgue measure and $B(x_i)$ is the interval $(x_i - 1, x_i + 1)$.

Now the norm $\|F\|_\varepsilon^*$, in the case $F = Lf$, is given by

$$\|F\|_\varepsilon^* = \left(\int_{\cup_{i=1}^k B_\varepsilon(x_i)} |f(x)|^p \frac{dx}{\varepsilon}\right)^{\frac{1}{p}},$$

where we have assumed that ε is small enough in such a way that the intervals $B_\varepsilon(x_i)$, $i = 1,...,k$, centered at the points x_i and radius ε, are pairwise disjoints. For a general function $F : B \subset I\!R^n \to I\!R^k$, the expression above has no meaning for $\|F\|_\varepsilon^*$ and we must use

$$\|F\|_\varepsilon^* = \left(\sum_{i=1}^{k} \int_{B_\varepsilon(x_i)} |f_i(x - x_i)|^p \; \frac{dx}{\varepsilon}\right)^{\frac{1}{p}}.$$

If we assume $F \in t^m$, $F = Lf$ and $T_m(F) = (p_1,...,p_k)$, we have

$$\left(\int_{B_\varepsilon(x_i)} |f(x) - p_i(x - x_i)|^p \; \frac{dx}{\varepsilon}\right)^{\frac{1}{p}} = o(\varepsilon^m),$$

as $\varepsilon \to 0$, $\;i = 1,...,k$. For obvious reasons we write $f^{(j)}(x_i)$ in the coefficients of the polynomial $p_i(x) = \sum_{j=1}^m \frac{f^{(j)}(x_i)}{j!}x^j$, and let $h \in \pi^{(m+1)k-1}$ be the Hermite interpolation polynomial which interpolates the data $f^{(j)}(x_i)$, $j = 0,1,...,m$, $i = 1,...,k$, i.e., $h = I(p_1,...,p_k)$ in the notation of the proof of Proposition 3.6.

Now fix $\varepsilon > 0$ and set $q_{\varepsilon,i} \in \pi^m, 1 \le i \le k$, for a polynomial which minimizes

$$\left(\int_{B_\varepsilon(x_i)} |f(x) - q(x)|^p \, \frac{dx}{\varepsilon} \right)^{\frac{1}{p}},$$

with $q \in \pi^m$. Then using Theorem 3.1 for the function $F(s) = f(x_i + s)$, we obtain $q_{\varepsilon,i}(x) \to p_i(x - x_i) = \sum_{j=1}^{m} \frac{f^{(j)}}{j!}(x_i)(x - x_i)^j$ as $\varepsilon \to 0$. Note that we have used the rather obvious fact that the polynomial $q_{\varepsilon,i}(.)$ is a solution of the above minimum problem if and only if the polynomial $p_{\varepsilon,i}(.) := q_{\varepsilon,i}(. + x_i)$ is a solution of the original minimum problem

$$\left(\int_{B_\varepsilon(x_i)} |f(x) - p(x - x_i)|^p \, \frac{dx}{\varepsilon} \right)^{\frac{1}{p}},$$

with $p \in \pi^m$. Now, the polynomial $h_\varepsilon = I(p_{\varepsilon,1}, ..., p_{\varepsilon,k}) \to h = I(p_1, ..., p_k)$ as $\varepsilon \to 0$. Thus we have obtained the Hermite polynomial which interpolates the data $f^{(j)}(x_i)$, $j = 0, ..., m$, $i = 1, ..., k$, as the limit of best approximation polynomials obtained minimizing on π^m the integrals on each neighborhood $B_\varepsilon(x_i)$ in an independent way. On the other hand, set $p_\varepsilon \in \pi^{(m+1)k-1}$ for a polynomial which minimizes

$$\left(\sum_{i=1}^{k} \int_{B_\varepsilon(x_i)} |f(x) - p(x)|^p \, \frac{dx}{\varepsilon} \right)^{\frac{1}{p}},$$

$p \in \pi^{(m+1)k-1}$. Then $p_\varepsilon \to h$. The proof is similar to that of Theorem 3.1. In fact,

$$\|(Lp_\varepsilon - Lh)^\varepsilon\|_\varepsilon = (\sum_{i=1}^{k} \int_{B_\varepsilon(x_i)} |p_\varepsilon(x) - h(x)|^p \, \frac{dx}{\varepsilon})^{\frac{1}{p}}$$

$$\le 2(\sum_{i=1}^{k} \int_{B_\varepsilon(x_i)} |f(x) - h(x)|^p \frac{dx}{\varepsilon})^{\frac{1}{p}} = 2\|L^\varepsilon(f - h)\|_\varepsilon = O(\varepsilon^m),$$

where we have used Proposition 3.6. Now use Proposition 3.2 with $P(t) = Lp_\varepsilon(t) - Lh(t)$.

4. The asymptotic behavior of the error

Let $\mathcal{A}$ be a subspace of polynomials $\Pi_k^m \subseteq \mathcal{A} \subseteq \Pi_k^l$ and let $F : B \subset I\!R^n \to I\!R^k$, be a Lebesgue measurable function. Set $P_\varepsilon \in \mathcal{A}$ for a polynomial which is a best approximation of the function F with the seminorm $\|F\|_\varepsilon^* = \|F^\varepsilon\|_\varepsilon$. Observe that $P_\varepsilon^\varepsilon$ is a polynomial in $\mathcal{A}^\varepsilon = \{P^\varepsilon : P \in \mathcal{A}\}$, which is a best approximation of the function F^ε with the seminorm $\|\cdot\|_\varepsilon$ within the class $\mathcal{A}^\varepsilon$, and we will also denote it by $P_{\mathcal{A}^\varepsilon,\varepsilon}(F^\varepsilon)$. We insist that $P_{\mathcal{A}^\varepsilon,\varepsilon}(F^\varepsilon)$ means, in our notation, a fixed best approximation polynomial and not a set of them.

Let $E_\varepsilon(F)$ be the error function $\varepsilon^{-m-1}(F^\varepsilon - P_\varepsilon^\varepsilon)$. Next, we will obtain an expression for the function $E_\varepsilon(F)$, which has its origin in [14] and [11].

Let F be in $t^{(m+1)}$ and set T_{m+1} for the Taylor polynomial of F of degree $m+1$; then by definition 3.1 we have $F^\varepsilon = T_{m+1}^\varepsilon + \varepsilon^{m+1} R_{m+1}^\varepsilon$, with $\|R_{m+1}^\varepsilon\|_\varepsilon = o(1)$, and $R_{m+1}(x) = \varepsilon^{-m-1}(F(x) - T_{m+1}(x))$. Moreover, observe that $\lambda P_{\mathcal{A}^\varepsilon,\varepsilon}(F^\varepsilon) = P_{\mathcal{A}^\varepsilon,\varepsilon}(\lambda F^\varepsilon)$ and ${T_m}^\varepsilon + P_{\mathcal{A}^\varepsilon,\varepsilon}(F^\varepsilon) = P_{\mathcal{A}^\varepsilon,\varepsilon}((T_m + F)^\varepsilon)$, where we have used that $T_m \in \mathcal{A}$. Then we have

Proposition 4.1. *Let F be a function in t^{m+1}, and $\Phi_{m+1} = T_{m+1} - T_m$. Then*

$$E_\varepsilon(F) = \Phi_{m+1} + R_{m+1}^\varepsilon - P_{\mathcal{A}^\varepsilon,\varepsilon}(\Phi_{m+1} + R_{m+1}^\varepsilon),$$

$\|R_{m+1}^\varepsilon\|_\varepsilon = o(1)$, *as* $\varepsilon \to 0$.

Proposition 4.1 is useful when $\mathcal{A}^\varepsilon = \mathcal{A}$ for every $\varepsilon > 0$. The case $\mathcal{A} = \Pi_k^m$ was considered in [14] and [11] for weighted L^p norms and in [4] for the Luxemburg norm. It is easy to find $\Pi_k^m \subset \mathcal{A} \subseteq \Pi_k^l$ and $\mathcal{A}^\varepsilon = \mathcal{A}$ for every $\varepsilon > 0$. The following result is relevant to this matter.

Theorem 4.1. *Let $\mathcal{A}$ be a subspace of polynomials such that $\Pi_k^m \subseteq \mathcal{A} \subseteq \Pi_k^{m+1}$. Then $\mathcal{A}^\varepsilon = \mathcal{A}$ for all $\varepsilon > 0$.*

Proof. We assume for simplicity $n = 1$. Clearly, there is a linear space $W \subset I\!R^k$ such that $\mathcal{A}_1 = \{Ax^{m+1} : A \in W\}$. Thus $\mathcal{A}_1^\varepsilon = \mathcal{A}_1$. Let H be in $\mathcal{A}^\varepsilon$, i.e., $H = Q^\varepsilon$ with $Q \in \mathcal{A}$. As $Q - T_m(Q) \in \mathcal{A}_1$ we have

$$Q^\varepsilon - T_m(Q^\varepsilon) = Q^\varepsilon - (T_m(Q))^\varepsilon \in \mathcal{A}_1^\varepsilon.$$

Thus $Q^\varepsilon - T_m(Q^\varepsilon) \in \mathcal{A}_1$ and there exists $V \in \mathcal{A}$ such that $Q^\varepsilon - T_m(Q^\varepsilon) = V - T_m(V)$. Therefore $Q^\varepsilon - V \in \mathcal{A}$, so $H \in \mathcal{A}$. We have proved that $\mathcal{A}^\varepsilon \subset \mathcal{A}$

for all $\varepsilon > 0$. Since $\mathcal{A}^{\frac{1}{\varepsilon}} \subset \mathcal{A}$ we get $\mathcal{A} = (\mathcal{A}^{\frac{1}{\varepsilon}})^{\varepsilon} \subset \mathcal{A}^{\varepsilon}$. In consequence, $\mathcal{A}^{\varepsilon} = \mathcal{A}$ for all $\varepsilon > 0$. □

Theorem 4.2. *Let F be in t^{m+1}, and $\mathcal{A}^{\varepsilon} = \mathcal{A}$ for every $\varepsilon > 0$. Then*
(a) $\|E_{\varepsilon}(F)\|_{\varepsilon} \to \|\Phi_{m+1} - P_{\mathcal{A},0}(\Phi_{m+1})\|_0$ as $\varepsilon \to 0$.
(b) $\|E_{\varepsilon}(F) - (\Phi_{m+1} - P_{\mathcal{A},0}(\Phi_{m+1}))\|_{\varepsilon} \to 0$ as $\varepsilon \to 0$ if $\|\cdot\|_0$ is a strictly convex norm.
We have denoted by $P_{\mathcal{A},0}(\Phi_{m+1})$ a polynomial in $\mathcal{A}$ which is a best approximation of Φ_{m+1} with respect to the norm $\|\cdot\|_0$.

Proof. The proof, in similar situations, is quite standard and we sketch it for completeness reasons. Let us begin with (a). By Proposition 4.1 we have, for any $P \in \mathcal{A}$,

$\|E_{\varepsilon}(F)\|_{\varepsilon} \leq \|\Phi_{m+1} + R^{\varepsilon}_{m+1} - P\|_{\varepsilon} = \|\Phi_{m+1} - P\|_{\varepsilon} + o(1) = \|\Phi_{m+1} - P\|_0 + o(1)$, as $\varepsilon \to 0$. Therefore

$$\overline{\lim_{\varepsilon \to 0}} \|E_{\varepsilon}(F)\|_{\varepsilon} \leq \|\Phi_{m+1} - P_{\mathcal{A},0}\|_0.$$

Let (ε_k) be a sequence converging to zero such that

$$\underline{\lim_{\varepsilon \to 0}} \|E_{\varepsilon}(F)\|_{\varepsilon} = \lim_{\varepsilon_k \to 0} \|E_{\varepsilon_k}(F)\|_{\varepsilon_k}.$$

Set $P_k = P_{\mathcal{A}^{\varepsilon_k},\varepsilon_k}(\Phi_{m+1} + R^{\varepsilon_k}_{m+1})$; then $\|P_k\|_{\varepsilon_k} \leq 2\|\Phi_{m+1} + R^{\varepsilon_k}_{m+1}\|_{\varepsilon_k} = 2\|\Phi_{m+1}\|_{\varepsilon_k} + o(1)$. By Proposition 3.1 we can select a convergent subsequence of P_k which is again denoted by P_k and then we have $\|\Phi_{m+1} - P_k\|_0 = \|\Phi_{m+1} - P_k\|_{\varepsilon_k} + o(1)$, as $\varepsilon_k \to 0$. Then $\|\Phi_{m+1} - P_{\mathcal{A},0}(\Phi_{m+1})\|_0 \leq \|\Phi_{m+1} - P_k\|_0 = \|\Phi_{m+1} - P_k\|_{\varepsilon_k} + o(1) = \|\Phi_{m+1} + R^{\varepsilon_k}_{m+1} - P_k\|_{\varepsilon_k} + o(1)$. Thus we have

$$\|\Phi_{m+1} - P_{\mathcal{A},0}(\Phi_{m+1})\|_0 \leq \underline{\lim_{\varepsilon_k \to 0}} \|E_{\varepsilon_k}(F)\|_{\varepsilon_k}.$$

To prove (b), consider any sequence $\varepsilon_k \to 0$ and select $P_k \in \mathcal{A}$ such that $\|E_{\varepsilon_k}(F)\|_{\varepsilon_k} = \|\Phi_{m+1} + R^{\varepsilon_k}_{m+1} - P_k\|_{\varepsilon_k}$. We will prove $P_k \to P_{\mathcal{A},0}(\Phi_{m+1})$, which implies (b). In fact we may assume, by taking subsequences if it is necessary, that $P_k \to P_0 \in \mathcal{A}$, as $\varepsilon_k \to 0$. Thus by (a) $\|\Phi_{m+1} - P_0\|_0 = \|\Phi_{m+1} - P_{\mathcal{A},0}(\Phi_{m+1})\|_0$. Henceforth $P_0 = P_{\mathcal{A},0}(\Phi_{m+1})$. □

As a first example of $\Pi^m_k \subseteq \mathcal{A} \subseteq \Pi^l_k$, consider the set $\mathcal{A} = \times_{i=1}^{k} \pi^{m_i}$, $m_i \geq m$, $i = 1, ..., k$. Then $\mathcal{A}^{\varepsilon} = \mathcal{A}$, for every $\varepsilon > 0$. The second example is the principal motivation of our study of the class of polynomials $\mathcal{A}$.

$$\mathcal{A}(l, k) = \mathcal{A}(l; x_1, ..., x_k) = \{Lp \ : p \in \pi^l\}, \tag{1}$$

where $Lp(s) = (p(x_1 + s), ..., p(x_k + s))$, and $-1 < x_1 < ... < x_k < 1$.

Proposition 4.2. *If* $0 < \varepsilon \neq 1$ *and* $k > 1$, *then* $\mathcal{A}^\varepsilon(l,k) \cap \mathcal{A}(l,k) = \mathcal{A}(0,k) = \{(c, ..., c) : c \in R\}$.

Proof. Let p, q be polynomials in π^l and let ε be a fixed positive number $\varepsilon \neq 1$, such that $Lp(s) = L^\varepsilon q(s)$, and we will assume w.l.o.g. that $k = 2$. From the identities $q(x_1 + \varepsilon s) = p(x_1 + s)$ and $q(x_2 + \varepsilon s) = p(x_2 + s)$ we obtain $q(x_1(1-\varepsilon)+\varepsilon s) = q(x_2(1-\varepsilon)+\varepsilon s)$, for all s. But if for a polynomial q we have $q(s) = q(s + c), c \neq 0$ for every s, then q is a constant. □

We cannot use Theorem 4.2 to study the error with $\mathcal{A} = \mathcal{A}(l,k)$, since Proposition 4.2 holds. The next condition on $\mathcal{A}$ will be significant in the future and will enable us to consider cases as $\mathcal{A}(l,k)$.

Condition 4.1. For $\Pi_k^m \subseteq \mathcal{A} \subseteq \Pi_k^l$, we assume that if $P \in \mathcal{A}$ and $T_{m+1}(P) = 0$, then $P = 0$.

Let $\mathcal{A}(l,k)$ be as in (1) and $l \geq k$; then we have $\Pi_k^m \subseteq \mathcal{A}(l,k) \subseteq \Pi_k^l$, where $(m+1)k - 1 \leq l \leq (m+2)k - 1$. Thus if $p^{(j)}(x_i) = 0$ for $0 \leq j \leq m+1$, $i = 1, ..., k$, then $p = 0$ and condition 4.1 holds.

The next theorem relates condition 4.1 with the property early used, $\mathcal{A}^\varepsilon = \mathcal{A}$ for all $\varepsilon > 0$.

Theorem 4.3. *Let* $\mathcal{A}$ *be a subspace of polynomials such that* $\Pi_k^m \subset \mathcal{A} \subset \Pi_k^l$, $\mathcal{A}^\varepsilon = \mathcal{A}$ *for every* $\varepsilon > 0$, *and condition 4.1 for* $\mathcal{A}$ *is in force. Then* $\mathcal{A} \subset \Pi_k^{m+1}$.

Proof. We denote by $\mathcal{A}_0$ the subspace $\{P \in \mathcal{A} : T_m(P) = 0\}$. Clearly $\mathcal{A}_0$ fulfills condition 4.1 and also $\mathcal{A}_0^\varepsilon = \mathcal{A}_0$, for every $\varepsilon > 0$; use $(T_m(P))^\varepsilon = (T_m(P^\varepsilon))$. We will prove $T_{m+1}(P) = P$ for every $P \in \mathcal{A}_0$. Then we will have the same property for every $P \in \mathcal{A}$, since for these polynomials we always have $P - T_m(P) \in \mathcal{A}_0$.

If $P \in \mathcal{A}_0$ is of the form $\sum_{m+1 \leq |\alpha| \leq l} A_\alpha x^\alpha$, with $A_\alpha \in I\!R^k$, we write

$$P^\varepsilon(x) = \varepsilon^{m+1}(T_{m+1}(P) + Q_\varepsilon(x)),\ Q_\varepsilon(x) = \sum_{m+1 < |\alpha| \leq l} A_\alpha \varepsilon^{|\alpha| - (m+1)} x^\alpha.$$

Since $P^\varepsilon \in \mathcal{A}_0$, we have $T_{m+1}(P) + Q_\varepsilon(x) \in \mathcal{A}_0$, for every $\varepsilon > 0$. But $Q_\varepsilon \to 0$ uniformly in B as $\varepsilon \to 0$. Then $T_{m+1}(P) \in \mathcal{A}_0$, which implies that the polynomial $R = P - T_{m+1}(P) \in \mathcal{A}_0$. Clearly $T_{m+1}(R) = 0$; then by condition 4.1, $P = T_{m+1}(P)$. □

We consider again the error function $E_\varepsilon(F) = \varepsilon^{-m-1}(F^\varepsilon - P_\varepsilon^\varepsilon)$, where $P_\varepsilon \in \mathcal{A}$ and $P_\varepsilon^\varepsilon = P_{\mathcal{A}^\varepsilon,\varepsilon}(F^\varepsilon)$. Set $G = F - T_m(F)$ and recall that $T_m(F) \in \mathcal{A}$. Then $E_\varepsilon(F) = E_\varepsilon(G)$. If $F \in t^{m+1}$ we have

$$E_\varepsilon(F) = \Phi_{m+1} - \varepsilon^{-m-1} P_{\mathcal{A}^\varepsilon,\varepsilon}(G^\varepsilon) + o(1), \tag{2}$$

as $\varepsilon \to 0$, and $\Phi_{m+1} = T_{m+1}(F) - T_m(F)$. The next theorem give us a useful expression for the error function $E_\varepsilon(F)$ as well as all we have to know for the polynomials $\{U_\varepsilon\}_{\varepsilon>0}$ and $\{Q_\varepsilon\}_{\varepsilon>0}$ used to describe it. With the notation $\mathcal{A}_0 = \{P \in \mathcal{A} : T_m(P) = 0\}$, observe that $\mathcal{A}$ is the direct sum $\Pi_k^m \bigoplus \mathcal{A}_0$.

Theorem 4.4. *Let F be a function in t^{m+1}, and assume condition 4.1 for $\mathcal{A}$. Set $\bar{P}_\varepsilon^\varepsilon = P_{\mathcal{A}^\varepsilon,\varepsilon}((F - T_m(F))^\varepsilon)$, with $\bar{P}_\varepsilon \in \mathcal{A}$, $U_\varepsilon = \bar{P}_\varepsilon - T_m(\bar{P}_\varepsilon)$, and $Q_\varepsilon = \varepsilon^{-(m+1)} T_m^\varepsilon(\bar{P}_\varepsilon)$. Then $U_\varepsilon \in \mathcal{A}_0$ and $Q_\varepsilon \in \Pi_k^m$ and*

$$E_\varepsilon(F) = \Phi_{m+1} - T_{m+1}(U_\varepsilon) - Q_\varepsilon + o(1), \tag{3}$$

as $\varepsilon \to 0$. Moreover the two families of polynomials $\{U_\varepsilon\}_{\varepsilon>0}$ and $\{Q_\varepsilon\}_{\varepsilon>0}$ are uniformly bounded in ε for a fixed norm $\|\cdot\|$.

The source of Theorem 4.4 is in [12].

Proof. By (2) we have

$$E_\varepsilon(F) = \Phi_{m+1} - \varepsilon^{-m-1}({U_\varepsilon}^\varepsilon + {T_m}^\varepsilon(\bar{P}_\varepsilon)) + o(1).$$

Or else, since $T_m(U_\varepsilon) = 0$,

$$E_\varepsilon(F) = \Phi_{m+1} - T_{m+1}(U_\varepsilon)^\varepsilon - \varepsilon^{-m-1} {T_m}^\varepsilon(\bar{P}_\varepsilon) + o(1),$$

as $\varepsilon \to 0$, which is (3). Since $\bar{P}_\varepsilon^\varepsilon = P_{\mathcal{A}^\varepsilon,\varepsilon}(G^\varepsilon)$, with $G = F - T_m(F)$, we have $\|\bar{P}_\varepsilon^\varepsilon\|_\varepsilon \le 2\|G^\varepsilon\|_\varepsilon \le 2\|T_{m+1}^\varepsilon(G)\|_\varepsilon + o(\varepsilon^{m+1}) = O(\varepsilon^{m+1})$. Therefore for a fixed norm $\|\cdot\|$ in $\mathbb{R}^k$, by Proposition 3.2 we have $\|\partial^\alpha \bar{P}_\varepsilon(0)\| = O(\varepsilon^{-|\alpha|}\|\bar{P}_\varepsilon^\varepsilon\|_\varepsilon) = O(\varepsilon^{m+1-|\alpha|})$, for $|\alpha| \le m+1$. Now observe that

$$\max_{|\alpha|\le m+1} \|\partial^\alpha P_\varepsilon(0)\|$$

is a norm on $\mathcal{A}$, where we are using that the subspace $\mathcal{A}$ fulfills condition 4.1. Thus $\bar{P}_\varepsilon$ and hence $U_\varepsilon = \bar{P}_\varepsilon - T_m(\bar{P}_\varepsilon)$ are uniformly bounded in $\varepsilon > 0$. To estimate the polynomials $Q_\varepsilon = \varepsilon^{-(m+1)} T_m^\varepsilon(\bar{P}_\varepsilon)$, we note that $\partial^\alpha Q_\varepsilon(0) = \varepsilon^{-(m+1)}\varepsilon^{|\alpha|}\partial^\alpha \bar{P}_\varepsilon(0)$, for $|\alpha| \le m$. Then $\|\partial^\alpha Q_\varepsilon(0)\| = O(1)$; recall that $Q_\varepsilon \in \Pi_k^m$, and $\max_{|\alpha|\le m} \|\partial^\alpha Q_\varepsilon(0)\|$ is a norm there. $\square$

Theorem 4.5. *Let F be in t^{m+1} and assume condition 4.1 for $\mathcal{A}$. Then $\|E_\varepsilon(F)\|_\varepsilon$ tends to*

$$\inf\{\|T_{m+1}(G-U)-Q\|_0 : U \in \mathcal{A}_0,\ Q \in \Pi_k^m\},$$

as $\varepsilon \to 0$, and $G = F - T_m(F)$.

Proof. We begin with the proof of the following inequality:

$$\overline{\lim_{\varepsilon\to 0}} \|E_\varepsilon(F)\|_\varepsilon \leq \inf_{U\in\mathcal{A}_0,\ Q\in\Pi_k^m} \|T_{m+1}(G-U)-Q\|_0.$$

Let $U \in \mathcal{A}_0$ and $Q \in \Pi_k^m$ be two arbitrary polynomials and set $S_\varepsilon \in \Pi_k^m$ with $S_\varepsilon^\varepsilon = \varepsilon^{m+1}Q$ and $P = U + S_\varepsilon \in \mathcal{A}$. Then

$$\|E_\varepsilon(F)\|_\varepsilon = \|E_\varepsilon(G)\|_\varepsilon = \varepsilon^{-m-1}\|G^\varepsilon - \bar{P}_{\mathcal{A}^\varepsilon,\varepsilon}(G^\varepsilon)\|_\varepsilon$$

$$\leq \varepsilon^{-(m+1)}\|G^\varepsilon - (U+S_\varepsilon)^\varepsilon\|_\varepsilon = \varepsilon^{-(m+1)}\|T_{m+1}^\varepsilon(G-U) - S_\varepsilon^\varepsilon\|_\varepsilon + o(1).$$

As $T_{m+1}^\varepsilon(G-U) = \varepsilon^{m+1}T_{m+1}(G-U)$, using the definition of the polynomial S_ε we have

$$\|E_\varepsilon(F)\|_\varepsilon \leq \|T_{m+1}(G-U)-Q\|_\varepsilon + o(1) \leq \|T_{m+1}(G-U)-Q\|_0 + o(1),$$

as $\varepsilon \to 0$, and the above inequality holds.
Now we will prove

$$\underline{\lim_{\varepsilon\to 0}} \|E_\varepsilon(F)\|_\varepsilon \geq \inf_{U\in\mathcal{A}_0,\ Q\in\Pi_k^m} \|T_{m+1}(G-U)-Q\|_0.$$

By Theorem 4.4 we select a sequence $\varepsilon_j \to 0$ in such a way that the following limits exist:

$$\|E_{\varepsilon_j}(F)\|_{\varepsilon_j} \longrightarrow \underline{\lim_{\varepsilon\to 0}} \|E_\varepsilon(F)\|_\varepsilon.$$

$$\lim_{j\to\infty} U_{\varepsilon_j} = \tilde{U}_0, \quad \tilde{U}_0 \in \mathcal{A}_0.$$

$$\lim_{j\to\infty} Q_{\varepsilon_j} = \tilde{Q}_0, \quad \tilde{Q}_0 \in \Pi_k^m.$$

Thus by (3) we have

$$\underline{\lim_{\varepsilon\to 0}} \|E_\varepsilon(F)\|_\varepsilon = \|\Phi_{m+1} - T_{m+1}(\tilde{U}_0) - \tilde{Q}_0\|_0$$

$$\geq \inf_{U\in\mathcal{A}_0,\ Q\in\Pi_k^m} \|T_{m+1}(G-U)-Q\|_0. \qquad \square$$

Proposition 4.3. *Let $\|\cdot\|_0$ be a strictly convex norm and assume that the subspace $\mathcal{A}$ fulfills condition 4.1. Then there exists a unique solution $(U,Q) \in \mathcal{A}_0 \times \Pi_k^m$ to the minimum problem in Theorem 4.5.*

Proof. Since $\|\cdot\|_0$ is a strictly convex norm it is enough to observe that if $T_{m+1}(U_1) + Q_1 = T_{m+1}(U_2) + Q_2$ with $(U_i, Q_i) \in \mathcal{A}_0 \times \Pi_k^m$, $i = 1, 2$, then $U_1 = U_2$ and $Q_1 = Q_2$. If $T_{m+1}(U_1 + Q_1) = T_{m+1}(U_2 + Q_2)$, then by condition 4.1, $U_1 + Q_1 = U_2 + Q_2$, but $\mathcal{A} = \mathcal{A}_0 \bigoplus \Pi_k^m$. □

Let us recall the following notation, see (2)-(3). For $F \in t^{m+1}$ we put $G = F - T_m(F)$ and $\Phi_{m+1} = T_{m+1}(F) - T_m(F) = T_{m+1}(G)$. For any $\bar{P}_\varepsilon \in \mathcal{A}$ such that $\bar{P}_\varepsilon^\varepsilon \in P_{\mathcal{A}^\varepsilon,\varepsilon}(G^\varepsilon)$, we set $U_\varepsilon = \bar{P}_\varepsilon - T_m(\bar{P}_\varepsilon)$ and $Q_\varepsilon = \varepsilon^{-(m+1)} T_m^\varepsilon(\bar{P}_\varepsilon)$.

Theorem 4.6. *Let F be in t^{m+1}, assume condition 4.1 for $\mathcal{A}$, and that the minimum problem in Theorem 4.5 has the unique solution $(U_0, Q_0) \in \mathcal{A}_0 \times \Pi_k^m$. Then $U_\varepsilon \to U_0$ and $Q_\varepsilon \to Q_0$ as $\varepsilon \to 0$. Moreover we have*

$$\|E_\varepsilon(F) - (T_{m+1}(G - U_0) - Q_0)\|_\varepsilon \longrightarrow 0 \quad as\ \varepsilon \to 0.$$

Proof. By Theorem 4.4, Theorem 4.5 and (3) any convergent subsequence of the net $\{(U_\varepsilon, Q_\varepsilon)\}_\varepsilon$ will converge to a solution of the minimum problem in Theorem 4.5. Thus if this solution is unique, then the whole net converges to it. □

5. The limit of best approximation polynomials

Let us recall that $P_{\mathcal{A},\varepsilon}(F)$ is a polynomial in $\mathcal{A}$, which minimizes $F - P$, $P \in \mathcal{A}$, with the seminorm given in definition 3.2. That is, $P_{\mathcal{A},\varepsilon}^\varepsilon(F)$ is a polynomial in $\mathcal{A}^\varepsilon$, which minimizes $F^\varepsilon - P$, $P \in \mathcal{A}^\varepsilon$, with the seminorm $\|\cdot\|_\varepsilon$; the latter polynomial may be denoted by $P_{\mathcal{A}^\varepsilon,\varepsilon}(F^\varepsilon)$. Thus $P_{\mathcal{A},\varepsilon}^\varepsilon(F) = P_{\mathcal{A}^\varepsilon,\varepsilon}(F^\varepsilon)$, and observe that the different seminorms used do not appear in this notation. Roughly speaking we may say that $P_{\mathcal{A},\varepsilon}(F)$ is a best approximation viewing the function on a small region, which is not the case of $P_{\mathcal{A}^\varepsilon,\varepsilon}(F)$, which can be the same polynomial for all $\varepsilon > 0$ and always considering a fixed region.

The main goal of this section will be to study the limit of $P_{\mathcal{A},\varepsilon}(F)$ as $\varepsilon \to 0$. If $F \in t^{m+1}$ and $G = F - T_m(F)$ it will be enough to consider $P_{\mathcal{A},\varepsilon}(G)$ as $\varepsilon \to 0$, since $P_{\mathcal{A},\varepsilon}(F) = P_{\mathcal{A},\varepsilon}(G) + T_m(F)$.

We set as before $P_\varepsilon = P_\varepsilon(G) = P_{\mathcal{A},\varepsilon}(G) = P_\varepsilon - T_m(P_\varepsilon) + T_m(P_\varepsilon) = U_\varepsilon + T_m(P_\varepsilon)$. Let F be in t^{m+1}; then $\|\partial^{(|\alpha|)} P_\varepsilon(0)\| \le O(\varepsilon^{m+1-|\alpha|})$ (see the proof of Theorem 4.4). Then $T_m(P_\varepsilon) \longrightarrow 0, \quad \varepsilon \to 0$. Thus $\lim_{\varepsilon\to 0} P_\varepsilon = \lim_{\varepsilon\to 0} U_\varepsilon = U_0$. From Proposition 4.3 and Theorem 4.6, this polynomial exists whenever $\|\cdot\|_0$ is a strictly convex norm. Then $\lim_{\varepsilon\to 0} P_{\mathcal{A},\varepsilon}(F) =$

$T_m(F) + \lim_{\varepsilon\to 0} U_\varepsilon = T_m(F) + U_0$, where U_0 together with Q_0 are the unique solution to the minimizing problem

$$\inf_{U\in\mathcal{A}_0,\ Q\in\Pi_k^m} \|T_{m+1}(G-U)-Q\|_0.$$

Thus if we set $P_0 = T_m(F) + U_0 \in \mathcal{A}$ for $\lim_{\varepsilon\to 0} P_\varepsilon$, then P_0, in $\mathcal{A}_F = \mathcal{A}_0 + T_m(F)$, will be the unique solution to the problem

$$\inf_{P\in\mathcal{A}_F,\ Q\in\Pi_k^m} \|T_{m+1}(F-P)-Q\|_0. \tag{4}$$

So, we have proved the following theorem.

Theorem 5.1. *Let F be in t^{m+1} and assume condition 4.1 for $\mathcal{A}$, and that the minimum problem in (4) has the unique solution $(P_0, Q_0) \in \mathcal{A}_F \times \Pi_k^m$, and denote by $P_{\mathcal{A},\varepsilon}(F)$ a polynomial in $\mathcal{A}$ which minimizes $\|F-P\|_\varepsilon^* = \|F^\varepsilon - P^\varepsilon\|_\varepsilon$, with $P \in \mathcal{A}$. Then $P_{\mathcal{A},\varepsilon}(F) \to P_0$, as $\varepsilon \to 0$.*

The presence of Q_0 in the statement of Theorem 5.1 is highly undesirable. An alternative statement can be done. We can set on Π_k^{m+1} the following:

$$\|P\|^0 = \inf_{Q\in\Pi_k^m} \|P-Q\|_0.$$

Then

$$\inf_{U\in\mathcal{A}_0,\ Q\in\Pi_k^m} \|T_{m+1}(G-U)-Q\|_0 = \inf_{U\in\mathcal{A}_0} \|T_{m+1}(G-U)\|^0, \tag{5}$$

and the polynomial P_0 of (4) is given by $P_0 = T_m(F) + U_0$, where U_0 is a solution of (5).

In some cases (5) can be explicitly calculated, for example in the classical L^p case (see [12]). As an example closed related to the one quoted above we set the following norms, which are independent of ε.

Example. By the same symbol, $\|\cdot\|$, we denote a monotone norm either on $\mathbb{R}^k$ or on the Lebesgue measurable functions $f_i : [-1,1] \longrightarrow \mathbb{R}$, and for $F = (f_1, \ldots, f_k)$ set $\|F\|_\varepsilon = \|F\| = \|(\|f_1\|, \ldots, \|f_k\|)\|$. Now if $P \in \Pi_k^m$, it is not difficult to see that

$$\|P\|^0 = \vartheta_{m+1}\|(P^{(m+1)}(0))\|, \quad \vartheta_{m+1} = \frac{1}{(m+1)!}\inf_{p\in\pi^m} \|p_{m+1} - p\|,$$

and $p_{m+1}(x) = x^{m+1}$.

Thus to obtain the solution $P_0 = T_m(F) + U_0$ to (5) we must find $U_0 \in \mathcal{A}_0$ such that

$$\|(G-U)^{(m+1)}(0)\| = \inf_{U \in \mathcal{A}_0} \|((G-U)^{m+1}(0))\|.$$

Observe that $\mathcal{A}_0$ is of the form $\{P^{(m+1)}(0)t^{m+1} : P \in \mathcal{A}\}$, which turns to be isomorphic to the subspace of $I\!R^k$ given by $\{P^{(m+1)}(0) : P \in \mathcal{A}\} = \{P^{(m+1)}(0) : P \in \mathcal{A}_0\}$.

References

1. P. Billingsley. "Convergence of Probability Measures", John Wiley & and Sons, Inc., (1968).
2. C.K. Chui, H. Diamond and L. Raphael. "On Best Data Approximation". Approx. Th. and its Appl. **1** (1984), 37-56.
3. C.K. Chui, H. Diamond and L. Raphael. "Best Local Approximation in several Variables". J. of Approx. Th. **40** (1984), 343-350.
4. H. Cuenya, M. Lorenzo, C. Rodríguez. "Weighted Inequalities and Applications to Best Local Approximation in Luxemburg norm". Analysis in Theory and Applications **20** (3), (2004), 265-280.
5. M. Cotlar and C. Sadosky. "On Quasi-homogeneous Bessel Potential Operators". Proceedings of Symposia in Pure Mathematics of the AMS, Vol **X**, (1967), 275-287.
6. C. K. Chui, O. Shisha, and P. W. Smith. "Best Local Approximation". J. of Approx. Theory **15** (1975), 371-381.
7. C. K. Chui, P. W. Smith and J.D. Ward. "Best L_2 approximation". J. of Approx. Th. **22** (1978), 254-261.
8. A.P. Calderón and A. Zygmund. "Local properties of solutions of elliptic partial differential equations". Studia Math. **20** (1961), 171-225.
9. S. Favier. "Convergence of Functions Averages in Orlicz Spaces". Numer. Funct. Anal. and Optimiz. **15** (3&4), (1994), 263-278.
10. W. Feller. "An Introduction to Probability and Its Applications". Volume **II** , (1966), John Wiley & and Sons, Inc., New York.
11. V. B. Headly and R.A. Kerman. "Best Local Approximation in $L^p(\mu)$". J. of Approx. Th. **62** (1990), 277-281.
12. M. Marano. "Mejor Aproximación Local". Phd Thesis (1986).
13. H. Maehly und Ch. Witzgall. "Tschebyscheff - Approximationen in kleinen Intervallen I. Approximation durch Polynome". Numerische Mathematik **2** (1960), 142-150.
14. R. Macías and F. Zó. "Weighted Best Local L^p Approximation". J. of Approx. Th. **42** (1984), 181-192.
15. C. Sadosky. "On some properties of a class of singular integrals". Studia Math. **27** (1966), 105-118.
16. J. L. Walsh. "On Approximation to an Analytic Function by Rational Functions of Best Approximation". Mathematische Zeitschrift **38** (1934), 163-176.

17. J. M. Wolfe. "Interpolation and Best L_p Local Approximation". J. of Approx. Th. **32** (1981), 96-102.
18. A. Zygmund. "Trigonometric series", Vols. **I, II**, third edition, Cambridge Mathematical Library, Cambridge University Press, 2002.